Automating Empathy

Automating Empathy

Decoding Technologies that

Gauge Intimate Life

ANDREW MCSTAY

OXFORD
UNIVERSITY PRESS

Oxford University Press is a department of the University of Oxford. It furthers
the University's objective of excellence in research, scholarship, and education
by publishing worldwide. Oxford is a registered trade mark of Oxford University
Press in the UK and certain other countries.

Published in the United States of America by Oxford University Press
198 Madison Avenue, New York, NY 10016, United States of America.

Library of Congress Control Number: 2023943870

ISBN 978-0-19-761555-3 (pbk.)
ISBN 978-0-19-761554-6 (hbk.)

DOI: 10.1093/oso/9780197615546.001.0001

Paperback printed by Marquis Book Printing, Canada
Hardback printed by Bridgeport National Bindery, Inc., United States of America

MIX
Paper from
responsible sources
FSC
www.fsc.org FSC® C103567

CONTENTS

ACKNOWLEDGEMENT

This book was made possible by generous support from the UK Economic and Social Research Council [grant number ES/T00696X/1] and Innovate UK [grant number TS/T019964/1]. Special mention once again goes to Vian for reading and commenting on drafts! Last, as anyone who has written a book knows, a book really is a product of many hands, not least a supportive editor, a meticulous copyeditor, and a careful proofreader. Thanks all!

Automating Empathy

On waking, Kimi's Garmin smartwatch displays the quality of her sleep: 7 hours and 54 minutes and a score of 81 from 100. 'Not bad', Kimi thinks, although she reflects that she could do with upping her deep sleep minutes. The smartwatch then begins the daily business of monitoring skin, heart rate, and voice for key emotional moments throughout the day. Coffee beckons, and Kimi calls to Alexa for the news. Amazon registers not just what she says, but how she says it. Awaiting the boiling moka pot, Kimi picks up her phone to scroll social media. It's election time and things are getting heated out there, with sentiment analyst Expert.ai opining that the race is closer than polls suggest. On to Instagram, her posts of political hope are logged.

With coffee in hand Kimi moves to the breakfast table to open her laptop and begin the morning emails. First, she calls to Alexa to play 'Upbeat Morning Music for Working'. Linked to Spotify, the recommender system responds, tracking mood type of Kimi's selections to provide moment-by-moment insights about Kimi to Spotify's advertising sponsors. Kimi begins composing her first email of the day. Working in sales, she has a tricky email to write to a prospective client. How to pitch the email? Long and detailed? Short and to the point? It's a media company, so should she go with exclamation marks to show excitement? Kimi opens Crystal that scrapes social media for insights on Neil, her procurement contact. It turns out that Neil is meticulous and critical. Neil will appreciate a detailed email. Definitely no smileys, or exclamations!

Automating Empathy. Andrew McStay, Oxford University Press. © Oxford University Press 2024.
DOI: 10.1093/oso/9780197615546.003.0001

And it's not 8 a.m. yet, nor has Kimi left home. When she does, Kimi makes her way to the train station. Passing through, live facial recognition is overlaid with computer vision that registers facial expressions and body pose, for security apparently. Exiting at Piccadilly Circus she looks at the ads, which look back at her, noting age, gender, and emotion expression. 'With my dark skin, do they see me?' she wonders. It's 9 a.m., and now at her desk on the busy sales floor, she is wearily resigned to having her sales calls scored for enthusiasm and positivity. She is perplexed by her average scores; Lisa, her ex-supervisor, used to praise her caring manner on the phone. It's now 11:00 a.m., and the Friday Feedback on the week's performance is in, provided in emoji. It's not been a good week: it's a downturned smiley. A meeting with her line manager Nasser beckons.

This book explores how it increasingly makes sense to speak of human measurement systems as 'feeling-into' everyday life. Such systems profile, judge, and interact with intimate dimensions of human life, attempting to perform and simulate empathy. Sometimes we will be aware of this, other times not. The 'automated' part refers to the premise that when systems have been built and deployed, data collection, processing, and actuation may occur with little human input beyond the person or group of people sensed. Automation processes are of course algorithmic, referring to the instructions and rules that computers follow when prompted. Whether these systems will in the long term do mostly social good or ill has yet to be determined, but as will become clear, this book has concerns. It is increasingly apparent that we will live 'with' and 'through' a variety of organisational practices and technologies based on use of data about the body to infer and interact with qualitative dimensions of human life.

At the most basic level the motivation to write this book is driven by a longstanding interest in the relationship between quantity and quality, and human entanglement with technology, especially systems that profile and make judgements about behaviour and subjectivity. To explore the appearance of such systems in everyday life, the book seeks to understand and place in context the emergence of automated empathy. This is the term employed here to identify systems that with limited human oversight are used to identify, quantify, respond to, or simulate affective states (such as emotions and cognitive states) and thereafter make judgements and decisions about individuals or groups of people.

Early note should be made on language in that 'automated empathy' functions as a meta-label to capture what has been variously termed 'affective computing', 'emotion AI', 'emotional AI', and 'empathic media', and related systems programmed to identify, quantify, judge, respond, and interact with emotions, affective states, cognitive states, attention, and intention. Preference for 'empathy' is due in part to its catch-all nature, but also its interactional properties. Although such systems do not possess subjectivity, nor have capacity for vicarious feeling, they also exceed what we usually mean by objects due to their programmed attempts to profile and interact with people on basis of emotions, moods, tastes, perspectives, and states such as sleepiness and intoxication, and even claims to preempt and track thought itself through brain physiology. The autonomous part of automated empathy of course refers to technical self-governance and systems capable of self-constituting processes. Yet it is the impression of separateness between people and process that motivates this book (Rahm and Kaun 2022). This separateness is experienced when a person or group is profiled without volition, when computational decisions occur without conscious input, when we are judged by unseen criteria, and that general sense of living in a technical environment that is not only made by somebody else, but for somebody else.

A key observation is that we increasingly 'live with' technologies that sense, learn, interpret, and provide feedback in ways that emulate human principles of empathy, especially cognitive accounts of empathy (Goldman 2008). This contrasts with mirror-based and mentalistic conceptions of empathy where there is vicarious experience (Gazzola et al. 2006; Oberman and Ramachandran 2009), which entails ability to understand the feelings of others because we too have experienced those feelings. How one understands empathy links to questions on the theory of mind. In relation to empathy this is dominated by two worldviews. The first is theorisation through observation, made up of concepts we employ that are 'adequate' to the task of getting us through everyday life. Opposite to warm and fuzzy sympathetic responses, these concepts equate to inferences, rules, principles, heuristics, laws, and explanations of the behaviour of things and people around us. The more intuitive and humanistic understanding of empathy denies that we come to understanding of others through 'theory', instead arguing that people use mental and other resources to approximate the experience of others and what it is like to be in the shoes of others (Maibom 2007). A key difference between the two views on empathy is that observational understanding of empathy is based on inference through public factors such as the behaviour of the body, situational context, and historical factors, where the past is a guide to automated decision-making in the present. The other more vicarious type, where one approximates to feel-into the other, involves private experience, where only those with similar experience may be able to react appropriately. In this account empathy is a necessarily human trait because it involves an imaginative leap to project oneself into the space that another person looks out from.

Automated empathy derives from the observational and theory-based view. As will be explored, the idea that a person's inner emotions or mental state can be accurately gauged from a person's face, gait, heart rate, tone of voice, or temperature is more than a little problematic. Still, ability to understand how a person is truly feeling is repeatedly claimed, even by companies that claim to be aware of the inherent problems with these technologies. Other claims to 'feel-into' human life are outright chilling, such as automation of lie detection, through video cameras and software proposed to track involuntary facial muscle movements (Shuster et al. 2021), in effect automating Dr. Cal Lightman, the fictional clinical psychologist who analyses microexpressions for evidence of deception in the Ekman-inspired crime drama *Lie to Me*, which ran on the Fox television network in the United States from 2009 to 2011.

Proponents, however, write headily of great promise and social progress through emotional intelligence and devices and mediated spaces that respond appropriately (Schuller 2017; Wiemer 2020). Claims include new modes of human understanding, communication, enhanced well-being, human–technology interaction, and novel ways of engaging with media content. Although sceptical, this book is also keen to avoid the bleak lure of the dystopian sublime that often characterises critique of deeply question-able use of technologies. Often figurative and poetic, dystopian accounts run risk of overamplification and distortion, failing to spot genuine harms for everyday life, missing opportunities for social and legal change, and by not allowing space for reenvisioning technologies (where appropriate) for more prosocial ends. Aversion to disasterism is not to suggest that all is well with human relations with technologies, and sensor and actuator-laden environments. All is not well and, as this book explores, the risks of automated empathy are significant. Yet, turning from the spectacle to the specific, it prefers to situate its commentary on emerging technology in context of the attributes and failings of technologies, deployment, govern-ance, and everyday human life.

THE EMERGENCE OF AUTOMATED EMPATHY

Despite association with artificial intelligence (AI), machine learning, and big data processing, the use of technology to attempt to infer emo-tion and psychological states has a technical history over 100 years old. As will be explored, this includes nondigital profiling of faces, heart rate, skin conductivity, the brain, respiration, and other biosignals. Moreover, AI itself, which involves machines that may simulate human perception, intelligence, behaviour, and abilities to achieve given objectives, also has a history of over 70 years (Turing 1950; McCarthy et al. 1955). It is notable in this regard that Alan Turing's 1950 paper on the question of whether machines may think was not a technical roadmap, but a paper about im-itation, mimicry, meaning, and authenticity, published in the leading philosophy journal *Mind*. Interest in the development of thought and human-like capacities in computers is at the very origins of AI. The idea that technologies may be empathic should raise eyebrows, as should the idea that by accessing physiology one can understand emotional life. As will be explored, the question of technological forms of empathy distils down to whether systems may function in ways appropriate to local and situation-specific social horizons. The test is and will continue to be not

whether such systems 'really' understand, but the extent of appropriateness in socially bounded contexts.

Accusations of snake oil rightfully abound about emotion recognition
and use of human-state measures to gauge subjectivity. The key problem
is that while physiology and emotion are connected, they are not one and
the same thing: emotion has social as well as physiological qualities. This
means that the sensing of bodies and oversimplistic judgement about what
expressions and behaviour signifies is very questionable. On the one hand,
this distils to stupidity and insensitivity to the context of an expression,
but in others it involves discrimination. The latter is illustrated by market-
leading systems such as Microsoft and Chinese company Face++ who were
both found to label black people with disproportionately negative types of
emotion such as anger, especially if there is ambiguity of what emotion
label to give to a facial expression (Rhue 2018). The consequence is scope
for discrimination in life-sensitive situations, such as education, work, security, and policing, to name but a few. As these technologies are being
deployed in important sites of social life, influential research institutes are
rightly flagging that there is a problem with both the technologies and how
they are being used. The AI Now Institute (2019), for example, highlighted
problematic scientific foundations and that AI companies should not be
allowed to deploy technologies that play a role in important decisions,
such as who is interviewed or hired for a job, the price of insurance, patient pain assessments, or student performance in school.

Claims of pseudoscience about inferring psychological and personality dispositions by assessing the body has a long history, reaching back
to the emergence of physiognomy (analysis of cranial and face shapes
to infer character) and pathognomy (analysis of temporary features and
expressions to identify emotive and episodic human states). Yet crude
usages of human measurement systems and highly questionable modern
activity should not be taken to be the end of the conversation, in part
because it risks positioning accuracy and bias as the core problem. This
book argues accuracy to be one problem, albeit a subpart of parental
problems regarding the larger question of organisational interest in profiling human subjectivity through systems based on, or overlapping with,

AI techniques. Although companies such as Amazon, Meta, Microsoft, and others continue to utilise caricatures of emotional life, the industry is internally aware that such simplistic approaches are problematic. It is caught between poles of expedience and attempts at efficacy. Staying with facial expressions for the moment, simplistic emotion labelling is pursued because facial expressions are relatively easy for object recognition systems to discern and categorise. More complex but meaningful accounts of emotion are harder and far less straightforward because they require more sources of data about a person and their situation to build confidence in what is being detected. Microsoft, for example, point out that 'only a very small number of behaviours are universally interpretable (and even those theories have been vigorously debated)' and that a 'hybrid dimensional-appraisal model will be the most useful approach' (McDuff and Czerwinski 2018: 79). The significance is that oversimplicity would be addressed by using more data about people, the situation they are in, and who they are with. Accuracy and whether the technologies work *are* important questions, but they are trumped by larger questions of whether we should be doing this in the first place.

When I first examined automated human measurement technologies in *Privacy and Philosophy* (McStay 2014) there were few meaningful commercial use cases of empathic technologies. With significant exception of profiling emotion on social media (and thereafter the profound implications for political life), much of the technology was used in neuromarketing labs and in-house research by companies. Just four years later, by the time *Emotional AI: The Rise of Empathic Media* (McStay 2018) was published, not only had the range of empathic technology start-ups exponentially increased, but they were competing for dominance. This is exemplified by Affectiva who aggressively promoted a freemium model of their software development kit to market services and train their emotion-sensing algorithms.

However, while the large technology firms (Apple, Amazon, Google, IBM, Intel, Facebook, and Microsoft) were developing emotion-recognition business services, they were still relatively quiet. This is changing. Today, Microsoft is building workplace analytics that detect stress and fatigue; Amazon's wristband, *Halo*, listens for emotion in voice; Garmin's and many other wearables monitor the body for emotion, stress, and well-being; social media companies such as Meta seek to profile and render facial expressions and other biometrics in virtual spaces; Spotify sells mood data to advertising conglomerates and has patents logged for use of biometric data; Ford, Honda, and many leading carmakers are equipping cars to track fatigue and profile emotion; ads in London's Piccadilly Circus to São Paulo's public transport system watch people to see if they like them; Lincolnshire police forces (in the United Kingdom) have tested emotion-enabled closed-circuit television (CCTV); Babylon Health traces emotion expressions so online doctors can respond appropriately to patients; Empath and others track callers and workers in call centres; Fujitsu and Softbank build assistive care robots that register emotion; AXA are building biometrics into insurance premiums; Intel offers schools cameras that monitor emotion and quality of engagement; and Mattel and Anki build toys that listen and watch for child emotion. Not all of the planned use cases will come to pass in the short-term, but the range of examples, when aligned with investment and development of computer vision and biometric technologies, at a minimum portends an ongoing use of digital quantification technologies in relation to qualitative dimensions of human life.

A closer look at a use case might help. A patent logged in 2021 by Spotify seeks to improve its speech-enabled recommender services by processing data from voice tone, speech content, background noise, user history, explicit indications of taste, profiles of Spotify user friends, and environmental data from background noise. As an overview, this involves personal data about self, place, and other people. Unpacked, it includes analysis of speech to determine (1) emotional state, gender, age, and accent; (2) where the Spotify user is and what they are doing (e.g., commuting on public transport, outdoors, at school [thus involving children], in a coffee shop, or in a park); and (3) who they are with (are they alone, in a small group, or at a party?). Location sensitivity goes further, stated to include 'sounds from vehicles on a street, other people talking, birds chirping, printers printing, and so on'. And that is solely in relation to the phone and the

Spotify thereon, but what of headphones and earbuds? These too are stated to have scope to track heart rate, blood pressure, and other personal vital signs over time, granting inference as to whether someone is listening, engaged with content, and where their spatial attention is directed.

This book does not aver that each detail in the patent will occur, especially as patents are often a wish list of interests, but the psycho-physiological direction and ambition of Spotify and other technology firms is clear. Notably, too, granular profiling is presented in the patent as a matter of *convenience*, to save the user having to 'tediously input answers to multiple queries in order for the system to identify the user's tastes', meaning automated empathy is positioned as a *benefit* to Spotify users. Although I for one would end my Spotify subscription if this level of data granularity came to pass, through this book's lens of hybrid ethics trends in interaction by means of the body are not outright rejected. Rather, the question is about what sorts of technologies people want to live with, on what terms, how do they affect the existing order of things, and at what cost do any benefits come?

Whether or not there are genuine social benefits, the ubiquity of image, face, voice, and biometric recognition technologies are increasing, assisted in some part by enhanced availability and decreasing costs. For example, the cost of training image-recognition systems (that include face expression recognition) in 2021 was 150 times less than in 2017 (The AI Index 2021). Policy makers, regulators, and technology standards bodies are now taking an interest in both growing prevalence of identity and emotion-recognition systems, although this is a very recent development. To illustrate, the European Union's 2018 General Data Protection Regulation (GDPR), which is regarded by most as a rigorous data protection package, makes no reference whatsoever to emotions. Similarly, a proposal for the ePrivacy regulation rarely mentions emotions. Only recitals 2 and 20 of the ePrivacy preamble mention emotions although, importantly, recital 2 defines them as highly sensitive (European Commission 2020). In the past I've argued there to be a clear lacuna in data protection about emotion (McStay 2016, 2018, 2019), but in 2021 this changed at international and regional levels. Internationally, the United Nations Human Rights

Council adopted a Resolution 48/4 titled 'Right to Privacy in the Digital Age', which recognised emotion-recognition technologies as an emergent concern. Recommendation 3 of this short document includes need for safeguards around 'emotional recognition' (UN General Assembly 2021: 4). Elsewhere, in 2021, the United Nations Convention on the Rights of the Child published 'General Comment 25'. This is significant because children are among the least able to protect themselves from negative features of technology uses but also historically underdiscussed in policy implements. Admittedly not a bombastic title, General Comment 25 moved to remedy lack of coverage by advancing and clarifying children's rights in respect to the digital environment, identifying emotion recognition as an emergent concern in the process.

In May 2023, members of the European Parliament endorsed the proposed AI Act, albeit with amendments. Seeking to address shortcomings of the GDPR, the AI Act highlights emotion. Prior to May 2023 the proposed Act indicated 'emotion recognition' as a priority issue by placing it at the top of the legislation in Article 1, although emotion recognition, along with other explicitly named AI systems, was later deleted from Article 1 and replaced in the amended version with 'certain AI systems'. Emotion recognition is explicitly named later in the revised AI Act though, with particularly sensitive applications being regraded from 'high risk' to 'unacceptable' in Europe. A risk-based piece of legislation, the AI Act differentiates between risks posed by different AI systems. First is 'minimal risk', meaning that there are no additional legal obligations to existing legislation. Next is 'limited risk', meaning that people should be notified of usage, potentially meaning that this would not give people possibility to opt-out (Malgieri and Ienca 2021).[1] 'High risk' AI systems are judged by harm to people's health, safety, fundamental rights or the environment. Use of these systems necessitate a certification mark before being allowed to be used in the EU. The criteria for risk include scope to contravene European rights, including values such as dignity, privacy, and nondiscrimination, but also more focused rights such as consumer protection, workers' rights, and rights of persons with disabilities. Some emotion recognition systems now fall into the high risk category due to it involving the special category of 'sensitive' personal data. The 'unacceptable' risk category protects against manipulation, exploitation, and social control practices. Now unacceptable and prohibited in the endorsed version of the AI Act is use of emotion recognition in law enforcement, border management, workplaces, and education institutions.

Mention should also be made of other prohibited AI in the proposed AI Act. For the most part prohibited AI does not involve emotion, but emotion is only one part of automated empathy. Automated empathy is also interested in other technological judgements about subjectivity and feeling-into individuals and groups of people. The AI Act prohibits 'subliminal techniques' that may distort behaviour and exploit vulnerabilities, which has application to neuro-technological instances of automated empathy discussed later in this book.

BOOK STRUCTURE

This book is sociological in orientation, drawing on philosophical, political, regulatory, and critical thought to discuss the significance of modern and emergent industrial technologies that simulate qualities of empathy. Where useful, it draws on empirical work conducted by the Emotional AI Lab, a collective I direct that studies the social impact of emotional AI and human-state measuring. Book chapters are divided into two sections. Section 1 covers definitions of key ideas underpinning assessment of automated empathy: unnatural mediation, hyperreal emotion, hybrid ethics, and the context imperative (Chapters 2–5). Section 2 discusses applications in specific settings and their implications (Chapters 5–11).

Chapter 2 of Section 1 introduces key historical and critical concerns that will recur throughout the book, especially regarding the artifice and imposition of a narrow understanding of emotional life. The chapter clusters diverse concerns under two headings: *unnatural mediation* and *mental integrity*. The first relates to oversimplicity, problems in method, and what will be phrased as a 'hyperreal' account of emotion. As will be explored, digital phenotyping and debatable theories of emotional life all impact on people through normalisation of a highly limited framing of emotion. This modern form of pathognomy, or inference of emotion and character through dynamic expressions, is not helped by opaque and demographically nonrepresentative data sets, already shown to have racist properties due to poor curation. *Mental integrity* relates to individual and collective mental states, privacy, cognitive freedom, and need for subject-centred protection. This recognises automated empathy interests in profiling and interacting with first-person perspective and the lifeworld of individuals and groups.

Chapter 3 addresses a recurrent criticism of emotion-recognition technologies (a key part of automated empathy) in the form of physiognomy, or the practice of assessing a person's character or personality from their outer appearance. As a human classificatory system that first boomed in popularity in the 1700s, it was inherently (if not intentionally) racist. Links with physiognomy are made in quality press articles (such as the *New York Times*) and in critical technology studies, but the question for this chapter is whether the criticism by physiognomy of modern emotion profiling is sensible. Does it follow that because emotional AI is akin to physiognomy in method, that emotional AI is also racist? This question is explored and answered through assessment of early physiognomy but also later and somewhat different Nazi-inspired later physiognomists. The chapter answers its original question but finds physiognomy to be highly influential on several relevant fields, including uncomfortable connections with early thinkers on empathy, phenomenology, and even critical theory. It also finds later physiognomy to be much less reductive in nature than early more commonly referred to physiognomy. Perversely, this later physiognomy tallies well with emergent and dynamic conceptions of automated empathy that admit of personal and contextual factors when making judgements about people.

Chapter 4 sets out the normative and philosophical orientation of the book. It starts with the observation that for many people modern life is now symbiotic with technologies that function by means of networks, and increasingly data about intimate aspects of human life. It proposes 'hybrid ethics' as a means of considering the terms of connectivity with sociotechnical systems. This may seem uncontroversial, but data ethics is seen here as less a matter of reserve and rights to be alone than interest in the terms of connectivity with systems and organisations. As will be explored, by beginning from an outlook based on human connection with data-rich technologies there is scope for charges of surveillance realism, overinterest in technology and usability, acceptance of dominance by major technology platforms, and seduction by marketing rhetoric. The hybrid view is argued to be mindful of these, instead purporting to rearrange and imagine alternative possibilities. At an applied level, the hybrid approach asks (1) what sorts of technologies do people want to live with; (2) on what terms; (3) how do they affect the existing order of things; (4) at what cost do the benefits of given technologies come; and (5) how can the relationship be better?

Chapter 5 considers the material and climate-based impact of automated empathy but assesses this issue in relation to Japanese thought about modernity, climate, and technology. Although the chapter recognises the extractivist lens for ecological concerns, which are argued to be deeply Heideggerian given intersection of concerns about industrialisation, cybernetics, and human being, the chapter progresses to examine climate-based considerations from a Japanese point of view, especially that of the Kyoto School. Beyond pluralistic value in diversifying and balancing moral views, especially given the hegemony that liberal and related thought has on global technology ethics, there is not only moral but rational merit in this. Japanese thought in and around the Kyoto School is valuable because it is ecological by default and innately oriented towards 'wholes' that encompass things, people, and environment. Argued in this chapter as a 'context imperative', the context imperative is argued to be a moral force that recognises the extractivist violence that dissociation and forgetfulness of the whole can do to the whole.

Having expanded upon key ideas that will resurface throughout the book, Section 2 contains Chapters 6–11 that assesses use and applications of automated empathy. The selection of use cases, issues, and examples are guided by several factors, including likelihood of becoming ubiquitous, harms, and in some instances positive uses.

Chapter 6 examines education because in many ways it is the social site most associated with empathy. In schools for example the best of teachers will gauge attention, child disposition, distraction, well-being, and depth of understanding, but what of technologies that pertain to help with this? Contemporary and emerging technologies exist in an education policy context where there is strong appetite for new systems and modes of analysing student performance. There is also keen interest in understanding the 'whole child', which involves inculcating traditional academic skills but also instilling of social and emotional aptitude for adult life. This has created interest in emotion, intention, and attention-based technologies for physical and online classrooms. The former involves cameras in classroom that read faces of students and teachers, and the online approaches that involve desktop computer cameras. Indeed, key companies such as Microsoft already have designs on profiling student emotion expressions in the metaverse, or whatever mixed reality ends up being named. Yet, what of the fundamental rights of the child, especially as recognised in the United Nations Convention on the Rights of the Child (UNCRC) and recent updates to it? Consideration of these rights and the 2021 General Comment 21 update to the UNCRC is especially valuable because this international governance certainly seeks to defend rights such as child privacy, but it also recognises the global right to good education. This includes not only wealthy countries with relatively low staff–student ratios and well-trained teachers but also regions that face economic hardships and profound challenges in educating the next generation. With potentially competing rights and interests, ethical answers are not straightforward. A hybrid approach to assessing a child's best interests assists by means of understanding the nature and terms of connection with technologies, those deploying them, and the conditions that need to be met for limited use of automated empathy in education.

Chapter 7 examines the significance of cars and an automobile sector that is increasingly feeling-into fatigue and the affective and emotional states of drivers and passengers. With policy backing in Europe, Ford, Porsche, Audi, Mercedes-Benz, and Toyota, for example, are all showing that this is one of the fastest growing applications of automated empathy.

Systems pertain to monitor fatigue, driver distraction, stress, anger, and frustration, but also to personalise the driver experience (such as by adjusting light, scent, sounds, and heating levels). While automobiles have long been an emblem of freedom, self-determination, and autonomy, this independence is being challenged by passive in-cabin profiling and generation of data that has scope to be used by third parties such as advertisers and insurance companies. A chapter with two different but related topics, the second part of this chapter on cars continues by exploring the overlap between autonomous driving and potential for affective states in cars themselves. Here, cars would simulate vulnerability to injury. Microsoft, for example, seeks to provide cars with something akin to an autonomic nervous system. A novel proposition, the principle is classically cybernetic involving goals, self-preservation, regulation, homeostasis, and behaviour. Yet, building safety into cars through appraisal is significant. It represents an artificially embodied and visceral approach to ethics. This visceral approach to safety and autonomous transport also provides for a unique take on the infamous trolley problem, in which an onlooker has the choice to save five people in danger of being hit by a vehicle by diverting the moving vehicle to kill one person. How is the problem affected by the trolley or vehicle 'feeling' vulnerable?

Chapter 8 examines automated empathy at work, focusing on use of technology to gauge well-being, temperament, behaviour, and performance. One of the more commonly studied instances of automated empathy (especially in employee recruitment), this chapter expands on current and emergent uses of automated empathy at work. Considering the call-centre workplace for example, it finds a bizarre and somewhat grim empathy game-playing at work, one in which workers are forced to game the demands of systems by caricaturing their own emotional expressions and empathy with callers. The chapter also considers visions for the future of work, again particularly those involving Microsoft. By means of analysis of patents, the impact of COVID-19, and biometric and mixed-reality technologies, the chapter sees a substantial shift towards immersion of expressive bodies in mediated communication and online work. It progresses to consider tentative interest from perhaps the most surprising of places: worker unions. Deeply hybrid, the latter part of the chapter considers worker-first approaches to automated empathy and the terms of connectivity required for a more 'sousveillant' (Mann et el. 2003) form of automated empathy at work.

Although much of the book considers emotion-based aspects of empathy, empathy is of course broader. In general, it is the ability to see and understand the perspective of another person. The perspectival part of empathy may involve understanding feelings and emotions, but it is also more than this: it is about all the factors that contribute to creating a first-person perspective. With empathy and perspective in mind, Chapter 9 considers nonclinical neurotechnologies. Of all the applications of automated empathy discussed in this book, consumer-level neurotechnologies and brain–computer interfaces may at first meeting seem the most outlandish. Yet it is an area of research that has received high levels of funding by companies such as Meta and Elon Musk's Neuralink. Moreover, well-funded organisations such as the US Defense Advanced Research Projects Agency (DARPA) have long explored the potential of immediate connections between the brain and computer sensors and actuators. Despite consumer-level brain–computer interfaces being potentially vaporware, which is technology that is promised but never arrives, the premise and implications nonetheless warrants critical attention. The chapter argues that brain–computer interfaces approach the apotheosis of automated empathy: the bridging of first-hand experiences and neurophysiology, or 'neurophenomenology' (Varela 1996). Indeed, with serious claims that human choice can be predicted before a person is aware of their own choice (due to delay between brain activity and conscious intention), this raises profound questions about [self] determination and relationships with technologies, each other, companies, and governments.

Chapter 10 asks a seemingly strange but increasingly important question: Is it morally acceptable for people to sell data generated by automated empathy profiling, such as data about emotions? With 'data trusts' and 'personal data store apps' receiving policy backing as a means of creating social value from data, this is a realistic proposition. The idea of 'infomediaries' is not new, reaching back to the original dot-com boom of the late 1990s/early 2000s (Hagel and Singer 1999). However, espe-cially given postboom intensification in profiling users of the web, the idea that people may control and even be paid directly for their data has stuck around. This has recently received increased technical and policy attention, especially given work on personal data store services by Sir Tim Berners-Lee in the form of the Solid platform. That an organisation such as the British Broadcasting Corporation (BBC) is developing services based on the Solid platform also makes it worthy of note, especially given the BBC's public service remit. They also invite moral questions, given urgency through increased use of biometrics and automated empathy. In reference to Michael Sandel's (2012) 'Moral Limits of Markets' thesis that states certain aspects of human life should be beyond monetisation and exchange, this chapter assesses personal data store apps that seek to grant people greater agency over personal data, potentially with view to making a profit from it. Instinctive aversion to monetisation of emotion and other intimate insights is complicated by the fact that they have long been commoditised, so why not have free-to-choose individuals making a share of the surplus value? Raising issues of seeing privacy and human rights through the prism of property, this chapter draws on survey work by the Emotional AI Lab. Providing a twist on what Sandel sees as corrupting market logics, the chapter finds tentative interest in scope to sell data to organisations that citizens believe to serve the public good in relation to health.

Chapter 11 concludes *Automating Empathy*. Very briefly recapping key arguments, the book ends with an assessment of what good looks like in relation to these technologies. The proposed answer is this: flip the entire premise of empathic systems. Here I argue that the premise of automated empathy, how people interact with it, and to an extent its governance, needs to be inverted, to begin with a worldview that is situated and local, com-plex rather than basic, prefers small to big, and promotes creativity over exploitation. Humanities-based, this is pluralistic, involving demarcated zones for playful and intimate interaction, based on self-labelling and local meaning-making.

Theory and Ethics

Hyperreal Emotion

The 2021 climate and political disaster satire movie, *Don't Look Up*, features an Apple-like launch event for a smartphone app called *LiiF* that claims to detect and help people manage their emotions. Against the bombastic backdrop of *The Creation of Adam* by Michelangelo, God hands Adam a handset. 'Let there by LiiF' is revealed on the smartphone screen, accompanied by the voiceover: 'Life, without the stress of living'.

As silver-haired CEO Sir Peter Isherwell prepares to enter the stage, the compere tells the audience to avoid sudden movements, coughing, negative facial expressions, or direct eye contact. Isherwell enters with three children to Apple-styled rapturous applause. Describing his life's work as an inexpressible need for a friend who would understand and soothe him, *Liif* is the answer, an app that 'is fully integrated into your every feeling and desire without you needing to say one single word'. Isherwell continues, saying 'If I feel', with the three children separately naming emotions 'Sad. Afraid. Alone', resuming saying 'the BASH 14.3 phone will set to the *Liif* setting, instantly senses my mood from blood pressure, heart beats'. Isherwell is then staged to be interrupted by the app as it cuts in, 'Your vital signs say you are sad, this will instantly cheer you up Peter'. The app screens a video of a puppy riding the back of a rooster, followed by offer of scheduling a therapy session. Now off-stage, one of the children approaches Isherwell to say, 'I love you, Peter'. Isherwell ignores her, coldly turning to ask employees in uber-marketing speak, 'Are you sure the video

Automating Empathy. Andrew McStay, Oxford University Press. © Oxford University Press 2024.
DOI: 10.1093/oso/9780197615546.003.0002

of the puppy on the rooster is optimising pre-pubescent sense memory consumer sector? I find the bird quite threatening.'

Don't Look Up depicts an inhumanity of industrial automated empathy, one whose public relations speaks of authentic connectivity yet whose corporate leaders and business models are depicted to be psychopathic, callous, unemotional, morally depraved, and hubristic. The movie also captures the functional gist of automated empathy, systems that profile and interact with intimate and subjective dimensions of human life, primarily by means of the body. This involves human profiling and use of extracted data that acts as proxy information for psychological, affective, emotion, and mood episodes. This chapter expands on the background to these practices, focusing on the nature of emotion, the history of emotion recognition technologies, and concerns and criticisms about these systems, especially regarding the artifice and imposition of a narrow understanding of emotional life. Highly deterministic, such datafied approaches really do see the face as the window to the soul, and that internal disposition can be confidently based on external information. As will be developed, there are fundamental problems with this view of how subjective and emotional life may be mediated. Yet, this view is influential, drawing power from a 'biosemiotic' view of emotions that turns to biology for truths of human-meaning-making rather than social and cultural worldviews.

This artifice is what will be argued to be *hyperreal*, where metaphors may stand in for reality and be presented as real. The risks are two-fold. First is that absence of ground-truth about what emotion is creates a vacuum for those who will define it in ways that suit them; and that these unnatural systems raise questions not only about efficacy and how emotions are represented, but also about potentially negative implications for the person emoting. Indeed, early use of automated empathy is already showing problems with racism. Second is that the increasing presence of automated empathy also invites concerns of a more psychological sort, at individual and collective levels. This will be phrased in terms of *mental integrity*, an expression that represents interest and social need for privacy, security, and uninfluenced freedom of thought.

BODY AND EMOTION

While emotions have been intensively studied in recent centuries, the philosopher of neuroscience and emotion Andrea Scarantino summarises that 'we are apparently not much closer on reaching consensus on what emotions are than we were in Ancient Greece' (2018: 37). Scholarship has generated diverse understandings and approaches to the question of emotion, including seeing emotion in the following terms:

- experiences of autonomic bodily changes (James 1884; Damasio 1999);
- excited states which motivate (such as to avoid danger), giving a purposive and causal nature to behaviour (Dewey 1894);
- behaviour patterns and dispositions of the body (Ryle 2000 [1949]);
- means of judging or appraising circumstances (Broad 1954; Arnold 1960);
- judging, perceiving value, and ethical reasoning (Nussbaum 2001);
- subcortical affect programs that primarily cause facial expressions (Tomkins 1962);
- evolutionary and biologically 'basic' programs that once initiated must be executed (Ekman 1977, 1989);
- culturally variable constructions built out of core affective states (Russell 2003; Barrett 2006); and
- emotion as fallacious premise (Scarantino 2012) because experiences of emotion are too heterogenous for emotion to be a useful label.

We will not resolve the nature of emotion here, but we can recognise it has experiential, intentional, communicative, judgemental, historical, cultural, and social qualities. Picking through these approaches, a reasonable if not somewhat stilted definition comes from Alan Cowen who observes emotions as 'internal states that arise following appraisals (evaluations) of interpersonal or intrapersonal events that are relevant to an individual's concerns' which in turn 'promote certain patterns of response' (Cowen et al. 2019). Unpacking this, emotions are argued to involve 'the dynamic unfolding of appraisals of the environment, expressive tendencies, the representation of bodily sensations, intentions and action tendencies, perceptual tendencies (e.g., seeing the world as unfair or worthy of reverence), and subjective feeling states'. This distils to questions of experience of the emotional state, expression, and attribution (labelling) by others.

Yet, rather than try to engage with accounts of emotion that embrace its richness, leading parts of the automated empathy industry remain loyal to simpler and more 'basic' accounts of emotion. This is the sort famously advanced by Paul Ekman whose corporate website, PaulEkman. com, showcases him as fifteenth among the most influential psychologists of the twenty-first century, *the* world-leading deception detection expert, co-discoverer of microexpressions, the inspiration behind crime drama television show *Lie to Me*, and as among the 100 most influential people in the world, according to *TIME* Magazine. For emotion-based automated empathy, no one has contributed more to the belief that emotion and other subjective conditions may be rendered as digital information, thereby making it capable of being extracted and put to work.

While this account of emotion is attractive because of its simplicity, core assumptions have repeatedly been shown to be highly flawed. I try here to avoid extensive recounting of the history of facial coding, basic emotions, and their failure, in part because I addressed it in depth in *Emotional AI* (McStay 2018). Others, especially Ruth Leys (2011), Lisa Feldman Barrett et al. (2019), and Luke Stark (2020), all offer deep and excellent critiques from diverse perspectives (Leys on material conceptions of emotion, Barrett on psychology, and Stark on the sociotechnical history of facial coding), all converging on the failings of oversimplistic, deductive taxonomies of basic emotions. Others provide useful context and support, situating interest in facial coding in relation to political, cultural, and affect-based theory (see Andrejevic 2013; Davies 2015; collection edited by Bösel and Wiemer 2020; Crawford 2021; Mantello et al. 2021; Stark and Hutson 2021; Cabitza et al. 2022).

In general terms, the nub of these critiques focusses on Ekman's view that facial expression of emotion is biologically based; facial expressions are universal across cultures; and that there is a causal link between facial expression and interior inside psychological phenomena (Ekman and Friesen 1969). The question is a simple one: Is facial behaviour reliably symptomatic of inner experience? Critically, the 'basic emotions' view suggests that emotions are reactive biological programs and that facial muscles function as a feedback system for emotional experience (Ekman and Friesen 1971, 1978). However, caution is required to not oversimplify critique because, contrary to what is sometimes assumed, the 'basic emotions' allows for cultural variation of these universal emotions and expressions through 'display rules' (Ekman and Friesen 1969). Writing in the Afterword to Darwin's *The Expression of the Emotions in Man and Animals*, Ekman is keenly aware of anthropological and social constructionist accounts that emphasise the role of culture in defining emotions, but nonetheless attacks Margaret Mead's argument that social behaviour is determined by culture. He tries to strike a balance, seeing emotions as having genetic and evolutionary programming, but also seeing this determinism as accommodating to social experience, display, feeling rules, and the situations that bring on emotions. However, despite admittance of variance in how people display emotion, he nonetheless argues that the expressions themselves are quite fixed because of the muscles required and that this fixity enables understanding both across generations and cultures (Ekman in Darwin 2009 [1872]: 387). Here emotion and correlate expression do not change, but the situational rules regarding when it is acceptable to express given emotions do change.

In theory, universality of emotion and its expression (albeit with regional and cultural adjustment) allow an analyst (be that a person or a computer) to 'reverse infer'. This means that if one identifies an emotion expression, one can deterministically trace back to the emotion episode that is taking place inside a person. This is assisted and advanced by the 'Facial Action Coding System' that Ekman developed with Wallace Friesen, which measures facial movement so that emotions may be labelled (Ekman and Friesen 1978). The taxonomy of expressions and muscles that cause those expressions has progressed to underpin digital versions of emotion recognition. This deductive system is based on Ekman's identification of seven facial expressions of primary emotion (joy, surprise, sadness, anger, fear, disgust, and contempt); three overall sentiments (positive, negative, and neutral), and advanced emotions (such as frustration and confusion), which are combinations of basic emotions. 'Action units' (or movements of individual muscles or groups of muscles) are further defined using an intensity scale ranging from A (minimum) to E (maximum) (McDuff 2014). Critical to his fame, Ekman also argues for the existence of 'microexpressions' that, in his account, are unguarded leaks of we *really* feel. Seen as involuntary emotion expressions between 1/3 and 1/25 of a second in duration (Li et al. 2015), these are deeply controversial, foremost due to their use to detect insincerity and lies. Moreover, despite the obvious criticism that a smile can be deliberate as well as spontaneous, proponents of facial coding and those who inspired it have long been aware of the difference (Duchenne 1990 [1862]; Ekman 1992), although the extent to which technology can differentiate is questionable (Kaur et al. 2022).

The approach is straightforward in principle: pay close attention to what the face does when it generates an expression said to be emotional by quantifying the relative positions of facial elements and changes in movement of these elements. With attention to eyebrows, cheeks, lips, nose wrinkling, and jaw movement, this (in theory) allows, for example, happiness to be labelled as a combination of cheek raises and lip corners pulled. As above, the movement of these action units can be quantified further in terms of intensity (do the cheeks and lips move a lot or just a little). Comparing real people's expressions to pregiven facial action labels (combinations of action units) is especially attractive to technologists with expertise in computer-enabled object recognition and image labelling. For them, this schema is attractive due to its simplicity and also that a computational worldview is already inherent in the basic emotions psychology itself because it sees emotions as programs. Application is relatively straightforward through image recognition of faces and classification of expressions by means of pregiven labels (again, combinations of action units). With computers this automates sensing of physiological change and reverse inference of an assumed emotion experience, also allowing this once human-only judgement to be done at scale. Among countless start-ups, major technology platforms use a variant of the basic emotions approach, including Meta (Facebook), Google, Apple, Amazon, and Microsoft.

Backed by Biology

In addition to the computational and mathematical qualities of emotion, the semiotic dimension is also important. Semiotics is the study of how meaning is communicated, which takes place through 'signs' (here being the combination of the expression and the emotion to which the expression is said to refer). In most semiotic studies elements of signs are based on arbitrary connections. Associations of blue for boys and pink for girls is arbitrary because they are based on learned cultural conventions, meaning there is no innate reason why these associations are so. Yet, the construction of the sign in the basic emotions worldview is *not* arbitrary. Instead, it is argued to be formed of a natural, physical connection between the signifier (such as facial expressions and other nonverbal gestures) and their meanings (the interior phenomena of emotions).

This can be unpacked in relation to the semiotics of Charles Peirce's (1867) field-defining three-part account of semiotics, based on icons, indexes, and symbols. Basic emotions have *iconic* qualities due to the claimed natural physical connection between the signifier (such as facial expressions and other nonverbal gestures) and their meanings (the interior phenomena of emotions). Also, basic emotions theory implicitly draws on *indexes* that have a factual relationship between the signifier and signified (just as a thermometer reading is a proxy for temperature). Explicitly rejected by the basic emotions worldview are *symbol*-based relationships, where the unguarded experience of an emotion would have no necessary correlate expression. This is because symbols are based on arbitrary relationships (like blue for boys, national flags, or traffic lights).

Indeed, because basic emotions are posited as biological programs, this lends well to *bio*semiotic considerations that (despite closeness of name) have fundamental differences from semiotics that studies symbol-based relationships established on arbitrary and socially contingent connections. Instead, the truth of signs is backed by nature and biology. Biosemiotics is the exploration of signs and meaning in a biological setting, encompassing human and nonhuman sign producers and sign receivers (Sebeok 1975). Schiller (2021) traces the historical connection between semiotics and Ekman, noting a friendship between Ekman and the biosemiotician Thomas Sebeok. Those familiar with Ekman will likely know him to champion Darwin's (2009 [1872]) phylogenetic, autonomic, and involuntary view that sees emotional expressions as public leaks of interiority, seen, for example, in animals when a cat's hair stands on end, or when animals puff out their chests to make themselves seem bigger than they are. This behaviour is key because it helps illustrate an ontology, not only featuring Ekman and basic emotions theory, but other modalities that involve tracking and gauging the body to infer some aspect of subjectivity. It is a non-anthropocentric worldview of signs in animal and human life, one that encompasses meaning making throughout the ecosphere, be these single-celled or complex organisms, where even unicellular creatures make 'judgements' based upon differences (e.g., temperature, unwarranted vibrations, being prodded, or the presence of nutrients; Damasio 2003: 30). Critically then, the biosemiotic position sees signification occurring without intentional communication. Deterministic, the consequence for a biosemiotic and basic approach to emotion is reverse inference of emotion as a fact of nature, that the face really is the window to the soul, and that sound judgement about internal disposition can be based on external information.

This may seem entirely sensible, especially when people are placed in their animalistic context. Yet even here there is evidence showing that apes (admittedly a close relation to people) can control and voluntarily regulate emotion expressions, including in the presence of audiences, which suggesting modulation of cognitive processes in relation to emotion and communication in social interaction (Kret et al. 2020). Other non-biosemiotic approaches to the history and make-up of emotions also provide enriching testimony for divestment from deductive universal basic emotions. This is especially so if one changes optics to consider more socially grounded and appraisal-led accounts to emotion, that recognise the existence of physical aspects to emotion, but also historical, social, ethnocentric, agreed, and co-created dimensions of emotion. 'Worry', for example, is a seemingly elemental and all-too-familiar emotion to many people, but the Machiguenga of Peru have no word for it (Smith 2015). In addition to criticism by diversity of understanding of emotion is the deductive nature of the method, especially in relation to what is missed. An analyst may well find instances of unwilled expressions, perhaps for example in joy at being reunited at airports (involving tears of joy and strong hard-to-label emotion expressions, as well as smiles), but what of the bubbling gamut of emotional life that makes up the rest of day-to-day emotional life? Important too is that the nature of emotions (or even their existence) is not all universally agreed or recognised by those with expert interest in what emotion is.

It is useful at this point to separate emotion science and research (that may involve face analysis) from simplistic industrial applications. Whereas critics diagnose the latter as pseudoscience, 'fake AI', and 'cheap AI' (Birhane 2021b), academic emotion research itself has developed richer and more sophisticated models of emotion to account for nuance in expression. This is not to argue for value in facial expression analysis, but to point to differences in complexity between industrial and academic work. Cowen et al. (2019), for example, recognise the coarseness and limitations of the Ekman basic emotions plus affective circumplex (comprising valence and arousal) approach. They note for example that feelings of sympathy or shame should not be taken as variants of sadness. Likewise, love, amusement, interest, or awe are not simply variants of happiness. Indirectly critiquing industrial applications, they also state that Ekman and Friesen (1971) overstated the significance of their findings regarding universal recognition of facial expressions and labelling of emotions, that people do not emote as caricatures, and that whether expressions indicate interior feelings, social intentions, or appraisals, remains problematic.

Work in face-based emotion research is keenly aware of its criticisms. Defending cross-cultural applications, Cowen et al. (2019) point out that while cross-cultural comparisons of single word labelling of emotions *are* flawed, when people are asked to link expressions to situations (a less linguistically oriented approach) results are improved (thus also factoring for human experience). Similarly, emotion researcher Alan Fridlund is widely cited as critical of Ekman's face-based work (Leys 2011). Yet, Fridlund's (2017) own work is face-based, albeit highlighting a more expansive and ecological understanding of emotion, rejecting basic emotion theory's quasi-reflexes, and seeing facial expressions as 'contingent signals of intent toward interactants within specific contexts of interaction' (2017: 77). This is highly context-based, involving rejection of reflexive and preprogrammed account of emotion in favour of one based on indications of intent in dynamic contexts. Other more nuanced approaches in emotion research are multimodal, arguing that awe is more readily sensed via voice than the face. Similarly, an emotion such as gratitude is difficult to convey (or measure) through face or voice modalities, but it is more readily readable in terms of ethnocentric tactile contact (Cowen et al. 2019).

The need for subtlety of criticism applies to the academic study of affective computing, which certainly may involve questionable assumptions about the nature of emotion. Often I have spoken at conferences where affective computing practitioners have presented advances on improvements in accuracy of emotion-as-object recognition, yet failed to comment on adequacy of the psychology. However, unlike industrial profiling of emotion, this is not uniform: many academic studies in affective computing *do* make use of self-reporting and/or compare the performance of technical systems with the ability of people to sense emotion (empathise) in others. Indeed, industry (Microsoft) and academic collaboration has also used self-reporting though user diaries to assess the validity of automated emotion recognition (Kaur et al. 2022). Thus, whereas applied and simplistic industrial profiling of emotion largely ignores questions of baseline truths and emotion thresholds, emotion science and affective computing may (but certainly not always) recognise fluctuations between people, and even for individuals, when gauged over periods of time (D'Mello et al.

2018). The argument being made at this stage is one of intellectual subtlety and that critics concerned about industrial emotion profiling should not confuse wilfully oversimplistic accounts of emotion with the wider fields of emotion research and affective computing. Just as basic emotions are rejected, we should be mindful of critical argument that is too basic. With wilful oversimplicity in mind, I turn now to introduce the idea of hyper-real emotion.

Hyperreal Emotion

A key concern connected with oversimplicity is reductionism, an argument repeatedly made by other critics. Although need for reductionism in computer science is recognised, AI ethics scholars of emotion also find that 'the conceptual models and proxy data for emotion in an AI system claim a descriptive power that is also a prescriptive power' (Stark and Hoey 2020: 8). This is important because the desire to simplify to play well with computer vision techniques has consequences, including the 'tendency towards a closed epistemology that values internal consistency in the face of evidence to the contrary' (Stark and Hutson 2021: 19). In other words, the problem with simplistic approaches is they are selective in what they look for: they only see what was is prescribed by a [basic] theory. Whereas the emotion labels attributed to physiological states are posited to have an indexical biosemiotic relationship to a more fundamental terrain of emotion, they do something else: they hide an ambiguity that is characterised by emotion being relational, incoherent, and less knowable. This is illustrated by a team featuring Lisa Feldman Barrett, well-known to be highly critical of emotion recognition. The question is a simple one, is there a reliable coherence between a facial expressions and experienced emotions? In work that tests relative merits of supervised (basic, prescriptive, deductive) and unsupervised (inductive) machine learning about emotion, they found 'above-chance classification accuracy in the supervised analyses' (Azari et al. 2020: 9). This better-than-chance finding could be seen as a weak vindication of the basic worldview. However, although the team found 'that there is some information related to some of the emotion category labels' (Ibid.), their examination of results from application of unsupervised learning showed a problem with this. With unsupervised learning being a process of inductively building clusters out of patterns in data, the team did not detect the same ground truths supposed by deductive theory due to the presence of other signals and patterns in the data. Whereas a 'basic' deductive approach stipulates what it wants to look for and finds it, in this case giving better-than-chance success, there also exists the question of how much is being missed about the nature of emotion responses and behaviour in the messiness of actual life, rather than that presupposed by narrow articulations. The significance is that if one stipulates criteria for supervised learning but does not challenge the merits of the psychological criteria and/or the emotion labels themselves, one may well find these patterns, but whether they represent a ground truth about the nature of emotion is a different matter.

Azari et al.'s (2020) work on supervised and unsupervised approaches to emotion is very reminiscent of Russell's (1994) critique of facial coding and basic emotions that underpin supervised learning of emotions. Reflecting on validity of methods and outcomes of Ekman and Friesen's (1971) famous study of cross-cultural emotion in New Guinea Highlands (that involved language translators, and behaviourally alert locals who were observed to be sensitive to subtle cues about how they should respond and react), Russell concludes by challenging Western ideas of universal emotion and expressions. Preempting Azari et al.'s (2020) findings through unsupervised learning, Russell (1994) observes that '[o]ne way to overcome the influence of such implicit assumptions is to emphasize alternative conceptualizations. And, I believe, the most interesting means to this end is to take seriously the conceptualizations (ethnotheories, cultural models) found in other cultures' (1994: 137). In other words, in both automated and nonautomated instances, the act of creating narrow emotion profiles is to impose a prescriptivist and somewhat territorial articulation of emotional life that has no ground truth, a hyperreal.

Despite the veneer of certainty provided by these approaches to emotion, dashboards, and attractive data visualisations provided to organisational users of emotion profiling systems, they promote a highly limited articulation of emotion, one that is especially problematic given the absence of lay or professional agreement on what the nature of emotion is (McStay 2018). This creates an imbalance, one in which language, number, and pictorial representations of emotion, create common understanding for something whose existence is in question (a basic thing to which its representation refers). This is deeply hyperreal, involving Jean Baudrillard's (1993 [1976]) idea that metaphors may stand in for reality, and may be presented as real. Indeed, the risk is not just that prescriptions are mistaken for the real (or at least recognition of the deep ambiguity of emotion), but that the hyperreal becomes reality. Admittedly Baudrillard may not be as intellectually fashionable as he once was, but in the case of emotion his concept of hyperreality is potent given that alluring data visualisation and big data may mask the truth that what emotion is remains debated, unclear, and likely beyond the reach of positivist truths. The difficulty of distinguishing between reality and its simulation is that while we might recognise shortcomings in simulation, there is no ground-truth to take recourse to. As Baudrillard puts it, 'It is only here [the hyperreal] that theories and practices, themselves floating and indeterminate, can reach the real and beat it to death' (1993 [1976]: 3).

Long associated with existential concern regarding industrially prescribed notions of reality, capitalism, perversion of social meaning, and ensnaring of interiority in commodity sign-production (advertising), those with a background in philosophies of media are likely to remember well Baudrillard's critique of the hyperreal, popular in the 1980s and 1990s. Applied here, industrially prescribed [basic] emotion risks conformity to a hyperreal articulation of emotion. However, unlike Baudrillard's hyperreal, human judgement through hyperreal emotion (which we can define as industrial prescription of emotion that takes place in the vacuum of absence of agreement on what emotion is) applies to diverse life contexts, far beyond (although including) marketing and the media industry. In addition to the social undesirability of having a limited palette of emotions prescribed by large technology companies, conformity and adherence to hyperreal emotion entails forgetfulness of the groundless nature of emotion.

Mulvin (2021) writes that proxies are products of suspended disbelief, which applies well to hyperreal emotion in that their expedience gives rise to what appears to be intentional avoidance of a modicum of critical thought regarding the meaningfulness of emotion labels. As will be explored, there are many political and representational problems with hyperreal accounts of emotion, but conformity to prescribed emotion also risks forgetfulness of the indeterminacy of emotion. Socially indeterminacy is valuable because emotion is a source for creativity, expression, and local meaning-making. With people usually having no need to theorise emotion, social defences of scepticism are down to hyperreal prescription. Although there exists a definitional vacuum, this is understandable: human empathy and 'emotion reading' are dispositional, meaning non-cognitive and non-procedural, meaning that people do not require named psychological methods before they begin interacting. Rather, the meaningful part of an exchange involving emotion comes from immersion in situations laden with agreed (and disagreed) meaning. In jargon, emotion is situated, embodied, distributed, and highly pretheoretical.

NUMBERING EMOTION: HISTORY AND PHYSIOLOGY

It is easy to think that hyperreal and automated accounts of emotion profiling are new, but Otniel Dror's (1999, 2001) chronicling of experimental interest in the physiology of emotion shows otherwise. The goal of this predigital and proxy-based interest was to standardise emotion through creation and use of technologies to visualise and represent emotions in tables, charts, and curves. Indeed, evident in the emotion taxonomies and photographic plates of Duchenne (1990 [1862]) the endeavour to mathematise emotion has long harnessed cutting-edge sensing technologies (electrodes) and modes of representation (photography) (McStay 2018). Critical, Dror (1999) describes highly suspect science, selection of research subjects who could be relied upon to display emotions on demand, and that work on using technology to number emotion took on racial properties. The latter will be addressed in the next chapter, but it is the techniques that are central here, pre-digital tracking of emotion by measuring temperature differentials, changes in heartbeat, blood pressure, breathing, conductivity of the skin, brain activity measures, facial coding, amongst other signals. All are present in automated empathy today but have their roots in the 1800s. For example, Dror (1999) traces laboratory-based accounts of emotion back to 1865 and Claude Bernard's account of the heart as both an object (pump) and psychological phenomenon. Dror (1999, 2001) also flags Angelo Mosso in the 1870s as inaugurating the history of physiology of emotion, seeing emotion, mental states, and physiology in parallel, taking the form of Mosso recording pulsation in the human cortex during mental activities.

The biosemiotic nature of automated empathy is supported by the initially surprising point that numbered emotion has partial origins in laboratory studies of animals. In Dror's (1999) assessment of the role of emotion and affect in early laboratory studies, emotion in animals was found to be a key element that influenced the physiological state of the animal patient, leading to early use of apparatus to track arterial pressure (which reveals heartbeat rate) and gastric measurements. An animal, for example, spending its first few days in a laboratory would be emotionally excited, therefore impacting on heart rate, red blood cell count, and body temperature, but once adjusted these would return to a more stable baseline. The worldview on emotion that this supported is also noteworthy, one based on purging aberrant emotion. Commenting on this early management of animals on laboratories, Dror notes that 'The new emotion-controlled "normal" represented nature's true physiological state' (1999: 217). Rephrased, the implication is that normality is controlled emotion, and emotionality means being out of control of either the body or situation. Reminiscent of the longstanding and somewhat Kantian dyad of rationality and irrational emotion, emotion was seen in negative terms of body failure and as a disturbance to be excluded. The outcome was that the laboratory was seen as both a source of emotional experience, but also the ideal place to study emotion because of scope to install controls and mitigating measures. Drawing on terminology from Stearns and Stearns (1985), this leads Dror (2001: 366) to observe that Western 'post-Victorian "emotionology" for people as well as animals was characterized by emotional restraint', this involving social rules regarding when the release of emotion was appropriate, such as in leisure, diaries, or therapy. Emphasising physiological dimensions (rather than spiritual and socio-cultural characteristics), aberrance, controlled normality, and emotions as leaky disturbances gave rise to a popular profusion of emotion and affect-gauging technologies, including the 'Lie Detector', 'Affectometer', 'Emotograph', 'Emotion-Meter', 'Stressometer', 'Psycho-Detecto-Meter', 'Ego-Meter', and 'Kiss-O-Meter' that each quantified hitherto qualitative phenomena (Dror 2001). While some may be relegated to historical curiosities, others are keenly relevant today. For example, the 'Emotion-Meter' of the 1940s was developed for cinemas to track pre-release spectator heartbeats and respiration though the ups and downs of filmic narratives, this preempting modern use of emotional AI in media research to optimise emotional experiences of ads, television, and movie content. Critically some of these early technologies were also seen as applicable *outside* of the scientific laboratory. As explored further in Chapter 7, Darrow's (1934) 'Reflexohmeter' preempted today's use of galvanic skin responses to measure in-cabin driver experience (and that of passengers) by some 90 years.

COMPUTING EMOTION

'Sentics' for Manfred Clynes are the communication of emotions (Clynes 1977). Clynes is important due to his influence on Rosalind Picard, who is synonymous with affective computing that connects numbered emotion, physiological views of emotion, technological approaches, and electric computation (see Picard 1997: 24–25, 28). Writing in his book *Sentics* of programmed emotion, keys, unlocking, the nervous system, biology, 'new mathematics', and cybernetic understanding of emotion, Clynes is curious because much of *Sentics* is about music, attempting to bridge the so-called two-cultures divide. His is the flipped side of a coin whose other side features key thinkers on affect, qualia, immanence, de-abstraction, sensation, and intensity (such as Sara Ahmed, Henri Bergson, Patricia Clough, Gilles Deleuze, John Dewey, William James, Brian Massumi, Sianne Ngai, and John Searle), who all play down the usual role of language to foreground immediacy and moments of experience. Although Clynes is very much interested in the codification of experience, a premise that philosophers of qualia reject, his interest in trying to make communication of expression more immediate chimes well with those that see affect in art in terms of embodiment, immanence, and sensation.

Important for accounts of industrial automated empathy is that emotion and their expressions are for Clynes indivisible. As with other biosemiotic accounts of emotion, this provides scope for reverse inference, meaning that an expression *does* (in Clynes' account) signal a definite experienced quality. Confidence is such that the expression of essential quality is not simply an ambiguous trace or after-event, but that emotion and expression are a functional system. Expressions of brain programs and their algorithms are thereafter enacted through diverse modes (faces, voices, gestures, and so on). Seeing his work as advancing that of Paul Ekman, Clynes argued that he had gauged inner shapes of emotion, allowing for *standardisation* of emotion. This is a scientific claim with industrial consequence, given scope to scale emotion profiling to products that are worldwide in nature (as marketed by Meta, Google, Apple, Amazon, and Microsoft). Ultimately this standardisation is expedient because emotional life is being reorganised around the limited units and measures that computer vision and emotional AI can process.

In the 1990s, via the work of Rosalind Picard, under the auspice of affective computing, this took the form of measuring heart fluctuations, skin conductance, muscle tension, pupil dilation, and facial muscles to assess how changes in the body relate to emotions (Picard 1997). Suchman (2007: 232) recounts that Rosalind Picard originally chose the name 'Affective Computing' for her project on datafied emotion because 'emotional computing tends to connote computers with an undesirable

reduction in rationality,' leading Picard to choose the term affective computing to 'denote computing that relates to, arises from, or deliberately
influences emotions' (also see Picard 1997: 280–281). One cannot help but
wonder whether the word 'affective' also provided a cover of seriousness
in the masculine context of MIT technologists, who at the time will have
seen AI in terms of cognition, symbols, representation, brain modelling,
processing, learning, rationality, and lessons from biological design and
situated experience. With risk of emotion studies in computing being seen
as a marginal, feminine, and imprecise topic, one can easily see how a revised project title was reached. This said, blurring of affect and emotion
has an older history, with the 'affects' of Silvan Tomkins (1962) referring
to less oblique emotional experiences, such as anger, fear, shame, disgust,
or joy. This leads Stenner (2020) to argue that Tomkins wanted a more
scientific term to dissociate his programs of work from common-sense
assumptions about emotions.

For Picard, use of computers meant that changes in emotion could be
sensed, they could be responded to by a system (a new premise), and a
system could consider information in relation to what it had learned before (another new premise) (Picard 1997). The 1990s were still early days,
but principles of sensing, actuation, and learning here argued to underpin
automated empathy were laid. This is due to pattern matching, in that a
computer is provided a standardised 'ground truth' of an affective state
and/or emotion expression. Whereafter the computer is programmed to
detect similar patterns in data (of words, images, and physiology), classify
the detected emotion expression, and react accordingly.

RECURRENT CONCERNS

Having introduced a working grasp of the psychological context of automated empathy and a genealogy of technologies that sense and interact
with subjective states (focusing on emotion), this chapter now distils
concerns into two principal concerns that will permeate subsequent
chapters: *mental integrity* and *unnatural mediation*.

Mental Integrity

Neuroethicist Andrea Lavazza (2018) defines mental integrity as an 'individual's mastery of his [sic] mental states and his [sic] brain data so that, without his [sic] consent, no one can read, spread, or alter such states and data in order to condition the individual in any way'. For Western thought, integrity-as-wholeness is found all the way back in Plato's (1987 [375 BC]) *Republic*, which sees the good life in terms of psychic harmony of appetite, reason, and emotion. It is one view based on intactness, absence of intrapsychic corruption by external interests, and ability to cope with environments that are taxing on mental integrity. Applied to the modern human-technical environment, which is already populated by systems designed to profile subjectivity, command attention, and influence people, there are risks. Mental integrity of self and society is not a priority for those who deploy these systems, and certainly not enough thought is given to the net result of complex connections with technologies and content therein. Indeed, regarding automated empathy itself, if these systems were judged by everyday standards of empathy, such system would be considered psychopathic and hardly conducive for mental integrity. Such systems can only pretend to care, relationships with them will be shallow and fake, and they are unable to form genuine emotional attachments, despite being marketing under the auspices of empathy and sensitivity to emotion. While most adults may be able to cope, suspend disbelief, and perhaps take gratification from such systems, what of children and otherwise psychologically vulnerable groups? And this is for the best of the systems: at worst, adversarial uses will seek to destabilise mental integrity and use automated empathy in ways that are antithetical to human wellbeing. While it is tempting to argue that systems claimed to have qualities of empathy should be held to moral standards associated with prosocial interpersonal empathy the problem is that this attributes moral agency to automated empathy systems. It is their developers and those who use deploy these systems that must be ethically and legally charged with responsibility to individuals and society.

With specific reference to neuroprosthesis (discussed in Chapter 9), mental integrity for Lavazza has technical as well as moral character, being concerned with functional limitations that should be incorporated into brain-based devices capable of interfering with mental integrity. This may be rephrased to be about concern with all intentions, plans, models, and processes, as well as devices and systems. Further, the concern is not just the brain and implants, but processes that use a broader set of automated empathy tools to unwantedly monitor and condition. The value of the expression 'mental integrity' is that it bridges privacy, security, and cognitive freedom, creating a useful catch-all term with which to consider many primary problems of automated empathy. These include rights to keep emotions and other subjective states to oneself, and to mental self-determination. In addition to being a useful shorthand for established rights, mental integrity is also appealing because it is subject-centred, seeing wholeness and integrity in terms of first-person perspective and the lifeworld of the individual. This is a subtle but significant point in that a social reality where people are effectively understood (or are at least predictable) from the inside-out is to be avoided at all costs. Such a view is some distance, for example, from legal language of the 'data subject'.

Sensitivity is required to the premise of first-person outlook because automated empathy does not 'understand' first-person outlooks. It uses proxies to try to represent first-person viewpoints and emotional dispositions through physiology and other information, in the case of the latter not least with social media (Bakir and McStay 2022). With proxies being those which have the authority to stand-in and represent some sort of truth (Mulvin 2021), the alignment between physiology and subjectivity is a broken but connected one. The two parts are attached but they do not function as supposed, where physiology would unproblematically reveal inner experience. Despite this broken connected status, this is enough to raise concern about privacy, security, agency, autonomy, and freedom of thought. The latter is not as obvious a concern as privacy but it is no less important given rights to not have thought manipulated, unwantedly surfaced, and not to be penalised for an emotion or thought (Scarff 2021). Importantly, mental integrity is here taken to embrace *collective* mental integrity. As argued in the pluralistically oriented Chapter 5 on Japanese ethical thought, individuals and all the richness of their unique outlooks must still be seen in context of the collectives that people belong to. As Nishida (2003 [1958]) argues, individuals and the deepest sense of subjectivity are expressions of society challenging and growing itself. The societal and collective dimension of mental integrity is crucial given challenges of hyperreal emotion, potential ubiquity, and perhaps most obviously in use of emotion profiling for political purposes (Bakir and McStay 2022). Consequently, while automated empathy involves intimate dimensions at a personal level, mental integrity is very much a collective psychological concern.

Unnatural Mediation

The second broad concern that this book has with automated empathy is unnatural mediation.

As introduced above, despite the simplistic allure of biosemiosis and basic emotions theory, where deep rules of nature would guide existence of emotions and their expressions, this is problematic on several fronts, including fundamental expert disagreement on the nature of emotion, the deductive nature of much emotion theory that only looks for what is presupposed, the communicative role of emotion, and context-specificity, among other problems. Emotion is not natural, because at least in part it is made or caused by humankind, not least the rich vernacular people in different life contexts use to articulate emotion. For automated empathy that mediates emotion and what are taken to be signals of subjective disposition, this too is unnatural. There are also profound political problems of mediation, especially regarding power relations, impacts on already marginalised groups, and that decisions made about people and life opportunities will be made in reference to problematic hyperreal insights.

Critique by unnatural mediation accordingly has two prongs: first is regarding industrial hyperreal articulation of emotion, but second due to political problems of mediation. Those with a background in media and communication studies will be long familiar with the idea that filters or processes which depict and mediate are prone to causing all sorts of problems. Historically this has for the most part had to do with how people portray other people and groups through media (albeit with consideration of the affordances and properties of media), but automated technologies add new interests. These are due in large part to their seeming independence from the messy and often ugly world of the cultural politics of how people are constituted and represented in automated technologies. This is especially the case regarding how women, people of colour, LBGTQIA+ people, people with disabilities, body shapes and types, and religious groups are [mis]represented in a variety of media systems. For automated empathy, limitations of computer sensing, biases in hardware and software, modern pathognomy (inference of emotion and character through dynamic expressions), opaque and demographically nonrepresentative data sets, digital phenotyping, and highly prescriptive and deductive theories of emotional life all contribute to unnatural mediation.

Despite a great deal of industry noise around AI bias, fairness, accountability, transparency, and goldrush for inclusive data sets and 'bias mitigation' tools (Zelevansky 2019), computer vision *still* has problems seeing darker skin as well as lighter skinned people. Worse, this distributed vision problem is not at all new because there is a history of bias in film-based photography being built for people with white and light-toned skin. Here, chemicals coating photographic film were not adequate to capture the diversity of darker skin tones. Should this seem a minor technical problem with an unfortunate outcome, more obviously egregious is that photography labs in the 1940s and 1950s even used an image of a white woman (called a 'Shirley card') as the basis for 'normal' (quite literally having the word 'normal' next to the picture; Caswell 2015). Calibration for whiteness is not simply a political criticism, but a historic fact of media, whether the means of representation be mechanical, chemical, or digital. On computer vision today, Wilson et al. (2019) find in their assessment of state-of-the-art object detection systems of two skin tone groupings that there are still disparities in the sensing of people. Similarly, Buolamwini and Gebru (2018) find that of all skin types darker-skinned females are the most misclassified group (being furthest from 'Shirley'), leading them to argue that 'these classification systems require urgent attention if commercial companies are to build genuinely fair, transparent and accountable facial analysis algorithms'. In 2022 Google signalled a move in the right direction by switching from the Fitzpatrick model of skin tone (that has six tones) to the Monk Skin Tone Scale (that has 10 tones) that is less biased towards lighter skin (Vincent 2022). The vision problem is not just skin tone though, with iPhone's facial recognition software having been found to have difficulty in distinguishing between different Chinese faces (Pinghui 2017). This is not helped by observation that people of Asian descent are among the underrepresented groups in Western training datasets (Gal 2020).

The predicament is not just vision based. In voice and speech recognition for example, Black speakers were twice as often misunderstood than White speakers in services by Amazon, Apple, Google, IBM, and Microsoft (Koenecke et al. 2020). This invites concern about need for code-switching to use technology services (Benjamin 2019) more easily, especially regarding not just what is said, but also how, in terms of emotion tone and change in prosody. Again, the political concern is that AI systems required to make judgements about people are being found to be calibrated to whiteness (Pugliese 2007), and what is otherwise a narrow slice of a global population, which occurs because social inequities are amplified through design practices that are blind to difference, poor training data, and inadequate algorithms (Noble 2018). Importantly, unnatural mediation is not just about who is affected by automated empathy, but the lifecycle of technologies and social contexts that support their development.

On the more focused question of emotion, Rhue, for example, found that emotion recognition systems assign more negative emotions to black men's faces than white men's faces (2018, 2019). Using a data set of 400 US National Basketball Association player photos from the 2016 to 2017 season, she found that emotional recognition software from Microsoft and Chinese company Face++ assigned black players more negative emotional scores on average, no matter how much they smiled. Black faces were always scored as angrier if there was any ambiguity about their facial expression, as compared to their white counterparts with similar expressions. This adds a perverse racial dimension to the hyperreal emotion discussion above, because for people with black faces to beat profiling systems, they must unnaturally increase positive expressions to reduce ambiguity and potentially negative interpretations by the technology. Consequently, although automated empathy currently replicates historical and social inequities in technical systems (Browne 2015; Mirzoeff 2020; Denton et al. 2021), ambiguity on the nature of emotion in automated empathy creates unique bias harms (Birhane 2021c).

There is also need for keen and careful treatment regarding whose emotions are conceived, especially regarding the ethnocentric character of emotion. Bonilla-Silva (2019), for example, discusses socially engendered emotions in racialised societies. This entails experiences that are inextricable from the societal shaping of the experiencing subject and groups that undergo these emotions, clearly making claims of emotion universality in hyperreal accounts problematic. The risk of standardised emotion (programmed into technology or otherwise) is that the 'dominant race's feelings are normalized' producing the status quo by which other groups are measured. A question of design justice (Costanza-Chock 2020) and the historic calibration of technologies of whiteness, the norms built into the designs of technologies by dominant groups cause difficulty and potentially material harm for others. Indeed, there may be intersectional characteristics, such as women of colour who may experience suppression of their emotions, such as shop workers when dealing with White customers, because of the nature of gender-race-class social positioning (Mirchandani 2003, cited in Bonilla-Silva 2019). With automated empathy increasingly featuring in workplaces, the potential problems are all too clear: face analytics that fail to see people properly, misclassification of expressions, and voice analytics geared that force workers from diverse backgrounds to code-switch, so they are not badly scored by automated empathy systems. This brings us to a key problem with bias-led critique that it does not ask the primary question: Should the monitoring be taking place at all? Indeed, critical focus on bias has helped to legitimise surveillance industries (Gangadharan and Niklas 2019) who are tasked with creating data sets and systems that are more inclusive, rather than being seriously confronted about the nature of the application. Industry legitimisation is perhaps only one part of the problem in that bias is an engineer's take on a social question, that is, one in which there is a deviation from a norm, that visible and known values may be adjusted and corrected, and that engineers are the right people to solve social problems as they are amplified through technology. Key is that while 'bias' is a technical problem that believes problems of representation can be corrected through optimisation and adjustment, those who see algorithms through the prism of 'systems of representation' will have a much larger domain of inquiry in mind regarding how inequity and injustices of all varieties have come to be.

CONCLUSION

The goal of this chapter has been to give a nonexhaustive sense of motivating ideas behind automated empathy and technologies that profile and interact with human emotion. Having recognised the oversimplistic (but appealing) deductive taxonomies of biosemiotic basic emotions, it framed the problem as one of a hyperreal nature. The risk is that metaphors for emotion might be presented as real, whereas the truth is that what emotion is remains debated, unclear, and potentially beyond the reach of positivist truths. In some situations, this will not matter, but in cases where judgements are made about people, such as at work or in education, the implications become more significant. Absence of an agreed ground truth as to what emotion is allows strategic interests and expedient understandings to fill the void. Closely related to hyperreal concerns are those of unnatural mediation, that asks questions about political problems of mediation and the precise reasons why people are misrepresented. For automated empathy, this involves hardware and software concerns, psychological and behavioural assumptions, and systems that have been found to be calibrated to whiteness and intolerant of diversity of expression.

However, 'bias' is seen here as an engineer's take on a social question: one based on fixing the problem rather than the larger question of whether such systems should be deployed in the first place. Concerns about automated empathy and technologies that sense and interact with subjective states are many, with this chapter clustering foremost concerns under the labels of mental integrity. Although acting as a catch-all for privacy, security, agency, freedom of thought, autonomy, and positive relations with others (be these people, organisations, technologies, or content), mental integrity is sensitive to first-person phenomenology (experience, disposition, and what may affect a person) of individuals, and the collective body that individuals belong to.

Assessing
the Physiognomic Critique

Chapter 2 flagged a range of critiques of emotion profiling and automated empathy, advancing these to be concerns of hyperreality, unnatural mediation, and mental integrity. Yet, it only briefly addressed concerns about physiognomy. Developed into a system for analysing people in the 1700s, physiognomy was the practice of using pregiven labels to identify a person's character or personality from their outer appearance, involving skull shapes, face geometries, and assumptions about race and culture. As early parts of this chapter note, the connection of physiognomy with emotional AI has become something of a given, certainly in critical writing about emotional AI. However, there is risk that a syllogistic mistake has been made. Distilled, critical writing of emotional AI argues:

Face-based physiognomy was racist (major premise)
Emotional AI is akin in method to early physiognomy (minor premise)
Face-based emotional AI is racist (conclusion)

Yet does it follow that because emotional AI is akin to physiognomy in method (in its mapping of regions of faces), that emotional AI is also racist? Or is this a mistaken syllogistic argument? One could argue that even if it is, given the potential for human rights abuses through emotional AI, who cares? Surely the ends justify the rhetoric, not least because

Automating Empathy. Andrew McStay, Oxford University Press. © Oxford University Press 2024.
DOI: 10.1093/oso/9780197615546.003.0003

the relationship between early physiognomy and emotional AI is a deeply ugly connection. Perhaps, but if the physiognomic connection *is* found to be syllogistic, might critical emphasis on a mischaracterisation divert attention from other potential harms of emotional AI and automated empathy more broadly?

Motivated by the physiognomic connection, this chapter assesses historical and philosophical factors behind physiognomy. Physiognomy refers to observation though cranial and face shapes to infer character, but lesser known is *pathognomy* that is analysis of temporary features (expressions) to identify emotive and episodic human states. Strictly speaking, although criticism of emotional AI focuses on the former, is it is the latter. Johann Kasper Lavater, perhaps the most well-known physiognomist after Aristotle, puts the distinction between physiognomy and pathognomy thus, 'The former shows what man is in general, the latter what he becomes at particular moments' (1792: 33).

The chapter finds that some of the physiognomy accusations stick, and others not. In its investigation into physiognomy of the 1700s and more recent physiognomy of the 1900s it identifies two forms of physiognomy: one is the *formal* sort that is associated with emotional AI and facial coding, involving pseudo-positivist taxonomies, quantification, and classification. The second is both more surprising and disturbing, being *humanist* in character. At this introductory stage of the chapter the conflating of physiognomy with humanism will be counterintuitive, but later physiognomy championed by Nazis (and others) rejected the pseudo-positivist and seemingly rationalist systems of physiognomy. Instead, humanist physiognomy was keenly alive to the errors of taxonomy, quantification, and biological reductionism, instead seeing selfhood in dynamic terms, reached through intuition and imagination. Alarming for a book about empathy (albeit of the simulated sort), this humanist physiognomy is inextricable from Germanic accounts of empathy that underpin lay conceptions of empathy, and arguably wider interpretivist and intuitionist thought. As will be developed, this leads to two outcomes: first is recognition that physiognomic ways of seeing and typification are pervasive; but second is more specific to automated empathy itself. This is that more recent approaches to emotion profiling and judgement of human disposition that eschew hyperreal labels in effort to engage with context and the dynamic nature of human life do not escape the physiognomic connection, they simply automate the humanist sort.

THE MODERN PHYSIOGNOMIC CONTEXT

An article for the *New York Times* distils the problem well. It states that facial recognition that functions in relation to emotion 'echo the same biological essentialism behind physiognomy' (Chinoy 2019). Unpacked, this means physiognomy is popularly understood as the preeminence of biology over culture, which in turn suggests dehumanisation, decline of individuality, and loss of the social. Seen this way, emotion is reduced to biometry and morphology (Galton 1901), the interest in body, shape, and form. Despite appeals to Enlightenment authority of quantity and positivism, it tacitly and explicitly discriminated against populations.

Diverse scholars have made similar connections (McStay 2018; Birhane 2021a; Crawford 2021; Stark and Hutson 2021). Each has critiqued emotional AI as being based a digital form of physiognomy, an AI-led form of judging a book by its cover (Stark and Hutson 2021: 13). Crawford for example suggests that physiognomy reached a high point during the eighteenth and nineteenth centuries in Johann Kasper Lavater's account (explored more fully later in the chapter) based on its appeal to objectivity and application of scientific methods, going on to cite Duchenne as another forerunner of Ekman's facial taxonomy. Similarly, Stark and Hutson (2021) claim Ekman-based emotional AI as physiognomic AI, pointing to an overlap between phrenology (the study of the skull as indicative of brain shape, mental faculties, and traits of character) and basic emotion theory, as they 'often assume a similar connection between a person's subjective emotional state and its outward expression' (2021: 27). Rightly, this critique notes appeals to science, rationalism, use of number, and classification, which contrasts with humanist accounts of emotion and subjectivity. This in turn may be associated with what is lived, dynamic, unbounded, emergent, nonquantifiable, and contextual. The problem with this reductionist and humanist split in relation to physiognomy is that most physiognomists are, strangely, humanists. As this chapter explores, this raises all sorts of implications for automated empathy, both in its current guise and the countenance of tomorrow's iterations.

EARLY PHYSIOGNOMY: SURFACE, NUMBER, AND INFERENCE

The influence of Johann Kaspar Lavater on automated empathy and its criticisms is hard to over-state. His *Physiognomische Fragmente zur Beförderung der Menschenkenntnis und Menschenliebe* (published between 1775 and 1778), hereafter *Physiognomic Fragments*, is important for introducing a worldview in which bodies were reduced to texts, where truths of human disposition could be interpreted through callipers and the language of number. It is with this understanding of physiognomy that emotional AI systems are frequently critically equated. Formally, the criticism is right: Lavater's physiognomy and the emotional AI industry markets understanding of human interiority by profiling exterior surfaces of the body. Moreover, Lavater sees it as capable of becoming a science, 'as any one of the sciences, mathematics excepted' (1792: 1).

Paradoxically, despite its racism, Lavater's goal for pseudo-positivist physiognomy was the promotion of generalised understanding and human love. The culture around early physiognomy is fascinating and recognisable. Daub (2018) details that physiognomy integrated itself into coffee table books and the bourgeois domestic sphere, interestingly noting that in addition to trying to classify others it also gave rise to anxious self-examination of one's own features against moral, intellectual, and animal profiles, but also zones of genius/idiocy, virtue/vice, and strength/weakness. Reminiscent of smartphone apps that achieve social virality through promise of foretelling ageing, or some other human insight 'achieved' through computer vision and large data sets, readers of physiognomy would engage in self-examination by benchmarking themselves to physiognomic standards distributed in books and photographs therein. The impact of physiognomy should then be understood as part of an assemblage of technologies that includes callipers, but also photographic, mechanical reproduction, and mass media. Thus, modern preoccupation with self-image and networked media, was again predated by the self-inspection of facial ridges and shapes some centuries before. Prescription, labelling of temperament, disposition, and personality through physiognomy translated into popular representation. For example, subjects of early photography and portraiture were careful in how they were depicted by artists, to ensure they were understood properly according to physiognomic principles. A hyperreal of sorts, physiognomic fictions blurred with reality to create groundless 'truths' played out in art and media.

Lavater argued that while physiognomy *is* based on an expert method, the act of inferring inner characteristics by countenance, or judging internal worth by external appearance, is common to all people. Pointing to the mutual inter-course of everyday life, for Lavater, using 'outward signs' to indicate 'some inherent quality' is the key to physiognomy (1792: 10). Physiognomy for Lavater is then simply the articulation and understanding of bodily and behavioural signification. Lavater sets his stall by asking:

[. . .] whether it be possible to explain the undeniable striking differences which exist between human faces and forms, not by obscure and confused conceptions, but by certain characters, signs, and expressions? Whether these signs can communicate the strength and weakness, health and sickness of the body; the folly and wisdom, the magnanimity and meanness, the virtue and vice of the mind? (Lavater 1792: 2).

For Lavater, physiognomy was both science and poetic interpretation, arguing that 'there must be a certain native analogy between the external varieties of the countenance and form, and the internal varieties of the mind. Anger renders the muscles protuberant; and shall not therefore an angry mind and protuberant muscles be considered as cause and effect?'

(Ibid.: 4). Arguing for self-evident truth of physiognomy, Lavater remarks that the shop seller will naturally and properly make judgements about the countenance and behaviour of those who enter 'his' shop. Honesty and sincerity, or whether a person has a 'pleasing or forbidding countenance', are then self-evident, leaving only the task of decoding and formalising the basis by which these judgements of human disposition are made. For Lavater this is not simply behaviour, but that 'The moral life of man particularly reveals itself in the lines, marks, and transitions of the countenance' (Ibid.: 29).

Should the pseudo-positivism of eighteenth and nineteenth century physiognomy seem distant and tenuous, consider that history repeats itself when modern machine learning researchers studying still facial images and criminality, make claims such as, '[. . .] the angle θ from nose tip to two mouth corners is on average 19.6% smaller for criminals than for non-criminals and has a larger variance' (Wu and Zhang 2016: 6). Debunking modern physiognomy and in assessing Wu and Zhang's claims, Agüera y Arcas et al. (2018) find numerous problems, not least differences in slight emotion expression between depicted criminals and noncriminals, that the police contexts in which photos were taken would likely lead to changes in expression, and that the 'ground truth' of the criminal look is based on human prejudice. Likewise, in an industrial context, the personality analytics company, *Faception*, also promises to find the criminal by their looks. There is good reason why Faception is routinely cited by critics (e.g., Stark and Hutson 2021), because the parallel between physiognomy and modern computer vision enabled classification is absolute. Moreover, while there is risk of using an outlying start-up firm to make a general point, that it has been able to do business (claiming customers in governments, retail, fintech, and robotics) points to interest from public-facing organisations. Faception's homepage is significant in this regard, claiming that 'We reveal personality from facial images' (Faception 2021), later claiming 'to know whether an anonymous individual is a potential terrorist, an aggressive person, or a potential criminal' and to be able to 'automatically reveal his personality, enabling security companies/agencies to more efficiently detect, focus and apprehend potential terrorists or criminals before they have the opportunity to do harm'.

Also reminiscent of the hubristic Dr. Cal Lightman (the fictional clinical psychologist who analyses microexpressions for evidence of deception in the Ekman inspired crime drama, *Lie to Me*), and the eugenics of Francis Galton (1885) who used images of men convicted of crimes to find 'pictorial statistics' that would justify essence of criminal faces, such uncritical belief is dangerous. Among problems with physiognomy and pathognomy (and phrenology) is that they reason backwards. This means biological behaviour is observed though cranial and face shapes to infer character (physiognomy), or analysis of temporary features (expressions) to identify emotive and episodic human states (pathognomy). Philosophically and ethically, physiognomy and acts of reverse inference in automated empathy butt against conceptions of the inner life. Certainly, there is problem in reading and labelling people, potentially with material consequences, but it should not be missed that both physiognomy and pathognomy are egregious because subjectivity is degraded (and missed) by methods both old and new. Such criticism has diverse roots in Western thought, including Socrates and Augustine of Hippo, where the inner life both exists, is valued, cannot be seen by outsiders, and should not be interfered with (Lavazza 2018).

The Hermeneutics of Face Science and Moral Semiotics

With hermeneutics being the theory and methodology of interpretation and determination of meaning, originally of the bible and literary texts, it can be applied to any sort of human actions, including the body. Physiognomy is premised on seeing and interpreting the body as a system of signs (Gray 2004), a point which connects well with Ekman's biosemiotic account of emotion explored in Chapter 2. After all, this is a key development of our age: ubiquitous usage of computers that 'see' and 'interpret' the world around them—recognising and classifying objects, animals, people, behaviours, and activities. This is what biometric forms of automated empathy do, they read and interpret people by prescribed means to make judgements about individuals and groups. The problem is twofold, with the first involving a more familiar connection between physiognomy and emotional AI profiling systems: Who is doing the interpreting, by what means, and with what effects? The second problem that will be developed in depth below is less familiar. Whereas physiognomy is critiqued on basis of biological reductionism, this is not quite right. Although the body and its behaviour and expressions are way-makers that indicate something of selfhood, these are means to point to a more authentic truth that exist beyond the body and [pseudo-] positivist method.

Gray (2004: 72) writes that for physiognomists including Lavater and Carl Gustav Carus, this 'science' also required a subjectivist 'artistry' to facilitate the decoding process. This has implications, regarding who is doing the decoding. Connecting with issues of gaze, power, and legitimacy, the analysts making judgements about race, identity, types, and human character were 'enlightened' white European men. The reason why physiognomy is so problematic is because reverse inference of *types* of people has, historically, involved racial profiling, and even worse, eugenics and genocidal search for purity. For critical exactness, the premise of physiognomy (which, to recap, is judgement of internal states through reverse inference of outward features) is not necessarily racist, but historically, physiognomies in both design and implementation have been racist and the literal antithesis of human flourishing. This is crystallised in the physiognomy of Hans Günther (1927), a Nazi, whose book, *The Racial Elements of European History*, is explicitly based on skull measurements and anthropometrics (anthropology based on physical measurement of people). Günther's *Racial Elements* is a very detailed set of profiles of European peoples in relation to physical countenance, yet what is assumed by these outward surfaces goes without detail or scrutiny. If this horrific connection between physiognomy and modern automated profiling does not lead to a direct ban on use in public spaces and to otherwise classify segments of populations, it amplifies need for transparency of method and questioning of the assumptions by which people are judged.

Written against the context of perceived decay, Günther's eugenic alarm about extinction through social equality is clear, arguing, 'The biologically untenable theories of the French Revolution (i.e., of the Ages of Enlightenment and of Rousseau) as to the "equality of all men" ended, as in France, by tearing down all over Europe the last barriers against race mixture' (Ibid.: 238). This was amplified in the industrial period, with its 'unrestrained race mixture' (Ibid.: 239). Praising of Francis Galton's eugenics, *Racial Elements* is an *avowedly* racist book, with Günther critiquing Spengler on 'overlooking the racial factors in historical phenomena' (1927: 199), involving a historical and global geo-political analysis based on heredity, blood, race, purity (especially of 'upper classes'), and identity.

Notable too given critiques of the universalising tendencies of modern emotional AI systems, for Günther the focus on race means there can never be a universal humankind, because races for Günther are predetermined by physiological and mental dispositions. One upshot of this for modern questions of automated empathy is that contemporary criticism of the universalising nature of these systems must also be open to the question of what assumptions will be made in design of heterogenous systems, that is, ones that are regionally grounded. Although automated empathy and emotional AI is critiqued for its universalising approaches to emotion and subjective life, there is arguably a greater physiognomic danger in diversifying automated empathy. The consequence is need for scrutiny of context-specific assumptions about regional populations, composition, skin and body shapes, identity, behavioural disposition, and emotion disposition and thresholds.

Although the eighteenth century physiognomy of Lavater was certainly racist by default in terms of its assumption and classifications, Günther's was actively founded on racist classification and aversion to social equality. Rejecting 'the common good', Günther expands that degeneration entails devotion 'to those with inferior hereditary qualities' (1927: 243). This sees the 'hereditary sound' (meaning those of 'unblemished' heredity and genetics) paying the way of the inferior and worthless, who in turn will use this assistance to increase birth rate. This leads Günther to praise programmes of sterilisation in the United States of criminals and the 'mentally diseased', chastising European lawmakers attempts to be 'just', although Günther elsewhere cites a 'maudlin sentimentalism' that has made America 'an asylum for the oppressed' (1927: 252). Günther instead championed those of 'Nordic blood' with their so-called creative and intuitive gifts, moral excellence, war-like nature, and 'being first into a fight', with Günther also arguing that racial mixing, social loss of light skin colouring, lowering of height, and changing of skull shape, produces decreased intelligence (or 'mental stock'), resulting in downfall and cultural decay. Critically, while this racial gaze was certainly biological, a true race science for Günther 'encompassed the whole "human Gestalt" (*Menschengestalt*) and not just partial physical or morphological features' (Varshizky 2021: 27). This wholeness is key to both physiognomy and its expressly racist off-shoots: accord between physical way-markers and psychic dimensions.

LATER PHYSIOGNOMY: BEYOND REDUCTIONISM

Seeing bodies as texts in the form of numbers and geometry, this *physiognomic hermeneutic gaze* plays another role. It treats the body (and the subject who may complain about this treatment) as a way-sign, a dumb marker for something truer that exists beyond the body. Here biologically essentialist critiques of modern emotional AI and automated empathy become weaker, because in later physiognomy the body and its morphology (shape and features) are seen as a set of patterns standing for a more authentic realm of meaning. Biosemiotic determinism is still present but features of the body are held to point to something metaphysical and spiritual in form. Thus, Günther's physiognomy involves biology and heredity, but it is not reducible to these due to the intuitionist factor in his and other related physiognomies. Taking hyperreal prescription regarding articulation of subjectivity through physiognomic systems to a horrific level, Günther's *Racial Elements* argued that race science should shape reality, to strengthen German consciousness of race. This moved science from *how* something works to the social question of *what for*. For Günther this latter question was a meaningful one to ask, even though it involved dimensions outside of the pseudo-positivism of race science itself.

Dynamic Humanist Physiognomy

Establishing a parallel between historic and contemporary reverse inference from the body to the person is necessary but easy. Yet, physiognomy has two prongs: one is *formal*, involving taxonomies, materialism, seeming positivism, quantification, and abstraction, all factors that feed into criticism of emotional AI as modern physiognomy; the second prong is *humanist*, receiving little to no attention in critiques of emotional AI. These could also be seen as modern versus antimodern, with the antimodern arguably the more sinister.

Humanist and antimodern prongs are important because the rejection of automated empathy and emotional AI on the basis that mathematical abstractions cannot capture the lived, dynamic, unbounded, non-quantifiable, and contextual aspects of emotion, is one also identified by humanist physiognomists. This is missed by a cursory examination of physiognomy. Spengler (1926) and Kassner (1919) for example are highly critical of taxonomy and abstract rationality, opting for a very different approach from Lavater's foundational accounts based on cataloguing and number. Despite the seemingly dehumanising character of number-based automated empathy, what physiognomy sought was precisely the opposite—its goal was a poetic understanding and soul-based conception of humankind. In many ways what was pursued was avoidance of spiritual catastrophism and a rebellion against rationalism, driven by anxiety over rapid social change, and aversion to liberalism, secularism, and industrialism. This concern and despair in Germany, especially among educated classes, was based on perceived technocracy, materialisation of life, and spiritual decay. The opposite of biological reductionism, spiritual anxiety in Fin de siècle Germany created conditions for popularity of 'life philosophy' (*Lebensphilosophie*), vitalism, romantic nature philosophy (*Naturphilosophie*) in biology, prevailing of *Gestalt* psychology, emergence of phenomenological psychology, and what in general was a rejection of positivism, mechanism and materialism (Varshizky 2021). Any view of physiognomy that solely emphasises anthropometry and phenotyping misses the point as it is a much more involved worldview.

Further, despite the perverse fascination of early physiognomic depictions (taxonomy, plates of skull shapes, face geometries, and expressions), the photographs belie recognition of the dynamic nature of selfhood and expression. Gray (2004: 197), for example, contrasts Lavater's early equivalence of static physical features with named character traits, to Rudolf Kassner's. Kassner's (1919) physiognomy differs by attempt to grasp the person in flux, involving principles of transition over stasis, motion, 'rifts' of being, and the analytical grasping of expressive drama. His exposition of need to understand identity in motion is reminiscent of process philosophies and phenomenology that emphasises need to understand that which occurs between momentary frozen instances, but also critiques of emotional AI that point to problems of using cameras and static expressions to capture experience. Kassner's physiognomy is more fluid than is generally supposed of either physiognomy or modern automated empathy, one that sees meaning in movement rather than stasis. Indeed, lessons can be learned in that while face-based critique of modern automated empathy typically fixates on the failure of capturing static images from the flow of behaviour, Kassner saw human being not as a static geographic map, but a moving and dynamic network. This is eminently visualisable today as the computer vision in facial recognition systems pinpoint regions of faces and dynamically tracks the ongoing movement of those regions. Indeed, broader organisational use of automated emotion recognition is increasingly aware of the issue of emotion variability and context. In discussion of emotion recognition and the workplace Kaur et al. (2022), a team featuring several Microsoft researchers, for example state that that computational assessment must factor for changing states and expressions in relation to physical and temporal aspects of a person's surroundings. While more dynamic conceptions of emotion profiling and automated empathy do not satisfactorily answer accuracy questions, they point the way to the trajectory of future iterations of these systems. These are based on granular real-time multi-modal profiling to admit of change and reactivity to place and circumstance. This then is less about capturing of static facial expressions but understanding change over time in relation to personal behavioural baselines.

Thus, the poetics of flux, flow, duration, concrescence, emergence and being, that qualitative writing is so fond of (and I count myself in this group) has unwanted colleagues in the bio-vitalism of later physiognomy and emergent forms of automated empathy. As Gray (2004: 199) highlights, the bridge for Kassner between the dynamic (in flux) subjectivity of others and the self is based on *imagination*, something entirely subjective, and some distance from static systems, abstract rationalising, and the universe of number. Imagination for Kassner serves to generate a unity, connecting with *communion*-based understanding of empathy (McStay 2018). This includes the physiognomy-inclined German thinkers discussed in this chapter, but also others who see empathy as a *social fact* that binds together communities through fellow-feeling. This sees empathy as not just an interpretive act (observation of behaviour, reactions to stimuli generated by others, and understanding of mental/emotional state of others), but also as a binding agent for a healthy co-operative society characterised by mutual understanding and awareness.[1]

Empathy and Physiognomy

For Kassner, imagination facilitates this physiognomically mediated empathic social reality, to render the object (person seen) as a subject. Phrased otherwise, *it is the transformation of quantity into quality*, which is quite different from neobehaviourist reading of physiognomy that reduces quality (emotions and disposition) to quantity (the plotting of expressions, movement, and behaviour). With origins in the Counter-Enlightenment that emerged out of opposition to eighteenth century Enlightenment rationalist doctrine, proponents insisted that matters of quality should not be lost in face of the rising tide of positivistic world-outlooks. Again, this

is deeply humanist, perhaps sitting at the heart of critique of emotional AI
and other automated technologies.

Others conceived physiognomic empathy in more mimetic terms.
Ludwig Clauss spoke of an empathy based on simultaneously maintaining
critical distance yet allowing oneself to participate in the type of person
being studied, articulating this in terms of studying racial objects by 'vis-
iting the world in which they live, and participating in their lives as much
as possible' (1929: 21). Clauss would eventually be expelled from the Nazi
party, in part because of the complexity of his racism. Whereas Günther
championed the Nordic races, arguing a clear understandable of material
formal differences between races, Clauss's physiognomy was grounded in
the language of empathy, flow, vitality, and phenomenology, as key means
of accessing human subjectivity. As Varshizky (2021) details, this involved
photography of bodies and faces to capture *sequences* of expressions and
actions, with view to revealing vitalist, dynamic, and emotional styles that
miraculate a racial body's movements. Directly connecting bodily and facial expressions with race, for Clauss features and outward behaviour were
the corporeal side of a latent inner racial soul (Varshizky 2021). Deeply
essentialist, Clauss' physiognomy was a *gestalt* that bridged body and soul,
putting physiognomy beyond the reach of quantitative methods. With
Clauss having been a student of Husserl, who argued that the body is the
medium through which spirit expresses itself, Clauss' racial perversion
of Husserl's psychology was founded on phenomenology and humanism
to escape classificatory logics, through movement over stasis, eventually
arguing for racial styles of being.

Already in this chapter physiognomy has been associated with some surprising areas of philosophy, but it reaches into core parts of the humanities
too. Johann Wolfgang von Goethe, who inspired and was related to a range
of protohumanists, was a self-stated physiognomist who saw the body as
a medium and outward features as an expression of inner substance. The
inner substance for Goethe was the soul, but as with German romanticism
more generally, this too was not a fixed thing, but also dynamic, changing,
and subject to renewal. Emphasising the less formal and more dynamic
branch of physiognomy, Goethe as physiognomist came to reject crude

abstract systems as a means of measuring and understanding human life (especially Lavater's). Care, then, is required in critique of automated empathy, in that physiognomy is highly diverse. It encompasses pseudo-positivism and abstract taxonomies as a means of profiling people, but at its core is interest in a more authentic form of human nature that transcends but is expressed through the body. Indeed, one of the key reasons that physiognomy is labelled a pseudoscience is that it overlaps with romanticism, where intuitionist faculties (nonverifiable assumptions) are required as a part of the physiognomy toolkit. This may seem a little strange, especially for readers used to a two-cultures diet, but many romantics were keenly interested in science (certainly Goethe, but others such as Schelling too). The difference between romantic understanding and that of the dedicated scientist was that subjective factors also played an important role in terms of appreciating that which is being studied, such as intuition of the vital forces behind evolving biological systems.

The same applies to early humanist accounts of empathy and Wilhelm Dilthey's *verstehen*, a principle that is core to sociology that seeks to interpret from the subject's position, rather than them being an object for the analyst. Similarly, a key originator of empathy, Theodor Lipps, argued 'Empathy is the fact that the antithesis between myself and the object disappears, or rather does not yet exist' (Lipps cited in Baldwin, 1940: 357). This is physiognomic: using reverse inference of observed behaviour to find a more real internal core which causes that behaviour. One consequence is that humanist empathy itself is open to the charge of reverse inference and physiognomy. Another issue is that interpretivism and reading of human symbols is, like automated empathy and emotional AI, that the seer, analyst, and the trained references they bring to bear on analysis goes unchallenged. Sociologists drawing on this tradition are reflexive and well-meaning, but the point stands: those drawing on phenomenological analysis, hermeneutics, and interpretivism, utilise a humanist physiognomic gaze. This sees in each of us a human interiority that structures how we perceive the world, and that this can be intuited by close analysis and interpretation of outward signs and information about context. Signs may vary, including for example speech in spoken and written forms, but the reverse inference premise is similar in that interpretive phenomenology traces outward behaviour and signs to make sense of how a person in each context sees things.

Oswald Spengler's physiognomy also draws on German neo-romantic ideas (especially Goethe and Nietzsche) of prerational knowledge, lived experience and intuition, to engage with the vitality of human being that animates the body. In *The Decline of the West* (1926), for example, its physiognomy is based on the idea of souls. This approach for Spengler escapes the gaze and logics of systems and geometry in favour of lived experience and immediacy. In jargon again reminiscent of phenomenology and process philosophies, Spengler remarks that, 'Becoming has no number. We can count, measure, dissect only the lifeless· and so much of the living as can be dissociated from livingness. Pure becoming, pure life, is in this sense incapable of being bounded' (1926: 95). The physiognomic critique of emotional AI and automated empathy must then be careful to avoid a two-cultures trap, because physiognomy is not at all defined by logics of number, behaviour, and observation. Again, for example, Spengler remarks that the 'Morphology of the organic, of history and life and all that hears the sign of direction and destiny, is called Physiognomic' (Ibid.: 100). This morphology is one of *style*, a physiognomy involving assessment of human fluidity and *disclosure* of soul and style through 'his statue, his bearing, his air, his stride, his way of speaking and writing', which is contrasted with 'what he says or writes' (Ibid.: 101).

The Phenomenology of Physiognomy

Physiognomists come in surprising guises, including figures such as Walter Benjamin whose urban wanderer 'was capable of reading occupation, character, origin, and lifestyle out of the faces of passers-by' (Benjamin, cited in Gray 2004: 219). This is significant, with Benjamin's gaze-based account of Paris being a hermeneutic gaze utterly predicated on physiognomy. Another well-known cultural critic and surprising candidate as a physiognomist is Theodor Adorno, who stated that 'knowledge of society which does not commence with the physiognomic view is poverty-stricken' (Adorno, 1976 [1969]: 33, also see Mele 2015). The point of these citations is to say that while it is right to connect automated empathy with physiognomy, as a *way of seeing* it is broad and pervasive in reach, oddly including sites of critique of physiognomy in the form of phenomenology (that seeks being and essences through phenomena) and semiotics (that analyses appearances for latent meanings).

A less philosophical assessment will help make the point about social pervasiveness, found in *Vogue UK* (Luckhurst 2020). In 2022 the UK's Home Secretary was Priti Patel, known for abrasiveness, bullying of her staff, inhospitable (many would argue inhuman) treatment of migrants, and a particular type of smile, designated by the UK press as a sneering 'smirk'. Interviewed in October 2020 on a flagship BBC news and current affairs program, *The Andrew Marr Show*, Andrew Marr finally snapped at her, 'I can't see why you're laughing.' The BBC subsequently apologised, conceding that what Marr interpreted as a 'laugh' was simply Patel's 'natural expression'. Without Patel's input, one cannot know whether she is intentionally communicating or inadvertently 'leaking' attitudes of smugness and superiority, but what is perhaps telling is the degree of confidence and conviction in the *intuitionist* physiognomic hermeneutic gaze. Despite AI-based anxiety about datafication of the body and parallel with physiognomic measurement, physiognomy runs much deeper in society.

This way of seeing is also apparent in Husserl's (1982 [1913]) 'pure phenomenology' that through an intuitionist lens also sees the body as the organ of the soul (Ibid.: 37). Remarking on the seeming strangeness of seeing a body walk down the street, rather than a person, by means of the 'eidetic reduction' Husserl points to need for understanding of human essences that have no physical extension yet *animate* physical bodies. Indeed, it is not outlandish to claim that nineteenth century humanities and the disciplines it initiated are predicated on the physiognomic way of seeing (also see Ginzburg 1989, Mele 2015). Much of this is simply to say that nondigital everyday stereotyping, perceptual framing, and discourses that informs appraisal of situations and others matter very much (Ahmed 2010, 2014). Applied to automate empathy interests, a consequence of seeing logics of reverse inference in people as well as automated empathy is that (1) any supposed dualism of humanistic and technological decision-making should be rejected; (2) attention to discrimination in technological decision-making should not let human decision-making pass with less critical scrutiny. Trite against when cast alongside Nazi physiognomy discussed in this chapter, one significance is that if reverse inference in automated empathy is to be rejected (and it should be, in situations even remotely sensitive), reverse inference also needs to be reexamined in everyday life too (again, especially in life-sensitive situations such as job interviews and policing where discrimination through physiognomy may occur). As Chapter 8 explores, automated empathy is used in screening of job candidates and, while critiques of automated human classification, bias-in/bias-out, and limitations of understanding all apply; human-facing screening is prone to the same physiognomic problem. Reverse inference, scrutinising of candidate behaviour to understand internal disposition (such as trustworthiness and sincerity), lack of transparency in how human interviewers have been 'trained', and what informs the ways of seeing by interviewing panels, all raise questions about generalised physiognomy. Indeed, one is reminded of Lavater (1792: 9) on analysis of facial countenance, who remarked, 'What king would choose a minister without examining his exterior, secretly at least, and to a certain extent?' Given that human inter-personal empathy first entails reading and judging the disposition of others through outward surfaces, the social practice of reverse inference is not easily discarded.

CONCLUSION

Critiques of emotion recognition technologies draw a parallel with the racism inherent in physiognomy, that linked human character with bodies, skull shapes, face geometries, race, and culture. These correspondent systems that connect outward and inward features and disposition, were either racist by default (Lavater) or avowedly racist. Yet, it is a stretch to say that modern machine-based judgement by appearance, or reverse inferences from facial and bodily behaviour to makes assumptions about a person is today innately racist. This chapter disagrees that racism in early physiognomy (or pathognomy) causes emotion recognition to be racist by default. However, the historic and method-based connection is strong enough to create urgency for need of analysis of the assumptions through which these systems see. The chapter finds that the chilling history of physiognomy should act as a spectre to illustrate the potential dangers of character, body, and expression typification. Developers and regulators of all varieties in the domain of automated empathy *must* recognise this history. It also means that, practically, reverse claims should also be closely checked regarding the ability of these technologies, along with prohibition of contexts of use.

The chapter also found that while physiognomy has materialist dimensions, such as mapping bodies and making inferences, the nature of physiognomy is such that the key problem is (admittedly strangely) intuitionist in nature. In physiognomy, the problem is not reducibility to the body, but that the body is a way-marker to a more authentic realm of meaning. As Nazi physiognomy shows, this can be horrific and perverse. Indeed, for a book interested in whether computers may simulate empathy, it did not expect to find intuitionist forms of empathy so deeply wedded to Nazi thought. One consequence of this more expansive understanding of physiognomy, that includes bodies as way-markers and intuitionist leaps to feel-into the other, is a that a two-cultures critique should be rejected. Humanistic critique based on observations of reductionism in physiognomy misses the point of physiognomy: it is about the leap to a more real realm of meaning, not reducibility of meaning to the body as behaviourism would have it.

The physiognomic critique as it stands today has done good work on grounding simplistic uses of emotion recognition in relation to Lavaterian physiognomy and pathognomy. Indeed, key parts of the automated empathy industry have backtracked on the validity of Lavaterian and basic emotions approaches in emotional AI. For example, responding to the reverse inference critique, Amazon AWS now post the following caveat:

The emotions that appear to be expressed on the face, and the confidence level in the determination. The API is only making a determination of the physical appearance of a person's face. It is not a determination of the person's internal emotional state and should not be used in such a way. For example, a person pretending to have a sad face might not be sad emotionally. (Amazon AWS 2023)

Yet, although this chapter recognises concern and disquiet around Lavater-like physiognomy in popular emotion recognition applications, it sees these as crude methods that will fall from view. Instead, it is the humanist physiognomy of Kassner and Spengler that is based on style and phenomenology of flow (that rejects quantity, preferring to assess what is in-between frozen moments) that is a portent of tomorrow's industry and social problem. This is an automated empathy that is dynamic in character, that rejects analyses based on context-free human expression, involving ongoing analysis of expression and behaviour in context of the who, what, and where of a person or group. While face-based analytics continue to be the public image of emotional AI and automated empathy, more passive and multimodal analyses (such as voice, core biometrics, big data, and online means) are better examples given scope for fluidic temporary assessment (what is occurring now, multimodally judged) and ongoing and longitudinal profiling. As physiognomy has historically shown, it is certainly connected with quantification of bodies, but the horror and error are one of quality, or what is assumed and argued of human nature through the process of reverse inference.

Hybrid Ethics

The liberal ethics that underpin much modern thought about the rights and wrongs of technologies were written with quills and ink. At a sociotechnical level, things have changed somewhat. A key change is that modern life involves symbiosis with interactive media technologies that function by means of networks and data about increasingly intimate aspects of human life. Large parts of this book for example were written against the backdrop of the COVID-19 pandemic that saw profound changes to human life, behaviour, and use of networked technologies for work and leisure. This deepened pre-existing social immersion in networked technologies. Does an increase in this imbrication necessarily involve a Faustian bargain, the paying of an unacceptably high price for diabolical favours? Should people not simply extract themselves from all arrangements and relationships with technologies, especially those used to simulate empathy?

This chapter sees that this is difficult, suggesting that questions about good and bad in relation to technology should start from the perspective that modern life is inextricable from technologies that function in relation to intimate and bodily data. A hybrid approach asks (1) what sorts of technologies do people want to live with; (2) on what terms; (3) how do they affect the existing order of things; (4) at what cost do the benefits of given technologies come; and (5) how can the relationship be better? This requires a view of autonomous systems not simply based on what these processes do to people, but one based on what people do with

Automating Empathy. Andrew McStay, Oxford University Press. © Oxford University Press 2024.
DOI: 10.1093/oso/9780197615546.003.0004

automated technologies (Pink et al. 2022). Moreover, while certainly new technologies may bring bad outcomes (with the wrongs of these needing diagnosis), what of rightness? In addition to harms and costs it is useful to consider not just what 'OK' is, but what good might look like. This chapter advances what the book posits as *hybrid ethics*, suggesting need to pay attention to the ethical implications of the enmeshing between people and autonomous systems that have qualities of emulated empathy. Perhaps controversially, it suggests that data ethics should *not* begin from the point of view of protecting a purely human sphere nor focus on the technologies alone. A hybrid view observes that separating people from technologies is a mistaken foundation for modern data ethics, because people are entangled in connective systems and platforms. Data ethics is seen here as less a matter of reserve and rights to be alone, but questions about the terms of connectivity with systems and organisations. In our case this is in context of novel automated empathy-based interactions that make use of data about bodies and subjective experiences. The hybrid stance does not exclude other approaches to ethics, but it does concentrate on the nature of connections between people and technologies, and what these relationships set in motion.

HYBRID ONTOLOGY

No discussion of hybridity and technology can take place without acknowledging Science and Technology Studies and the theorist, Bruno Latour. Latour's (1996, 2005) account of technology is an approach that sees objects as social and as worthy of respect as well as critical attention. The word 'social' simply refers to a set of connections encompassing technical items and processes, although Latour points to a more formal sense of representation. His discussion in *Politics of Nature* (2004) involves an extension of democracy or political ecology to nonhumans. The defence of this is somewhat convoluted but the means by how he arrives at this are more important than his proposed republic or parliament of things. This is done by means of disturbance of the proposition that only human actors have will and intention, while objects may only behave and be subject to known causal norms. For Law (2012 [1987]), along with others of an Actor Network Theory (ANT) disposition, this opening-up to the multitude involves the witnessing of heterogeneous elements assimilated into a network. This 'ontology' sees a plurality and coming together of social, economic, political, technical, natural, and scientific entities on the same terrain as each other. In *An Inquiry into Modes of Existence*, Latour (2013) affirms the need to span modes of existence and hybridise symbols/things, culture/nature, and the various ways in which these mutually inform and infuse each other to destabilise premises of absolute categorical distinctiveness. An influence of note on Latour is Gabriel Tarde's expanded sense of what is meant by 'society,' which is also multiscalar, including all sorts of nonhuman entities (be these bacterial, celestial, or machinic). As will be unpacked in relation to Japanese ethical thought in Chapter 5, the hybrid ontology involves both understanding and openness to the life of things. The direction of travel is likely already clear, to a place where ethics is about appropriate terms of hybridisation, one that prioritises mental integrity in context of nonhumans programmed to interact in more humanly intimate ways.

Surveillance Realism?

Adversarial, this proposition of hybridity and enmeshing could be read as an example of 'surveillance realism' where the existence of surveillance infrastructures limits the possibilities of imagining alternatives (Dencik and Cable 2017), but it is not. It could also be read as falling into the 'legitimacy trap' described in human-computer interface literature (Dourish 2019), where usability is prioritised over dignity, but this is not the case either. It could be read as accepting dominance of major technology platforms and legitimising the power of technology platforms (Whittaker 2021), but this too is not the point. Finally, it could be seen as seduction by 'industrial imaginaries' (Pink et al. 2022) where automated technologies are marketed as providing magical solutions to problems. These of course are self-referential situations, where the promoted technologies seem to make perfect sense if interpreted through the prism of worldview and claims presented by ads and marketing (Berg 2022). There are dangers though of the unwary academic accepting face-value claims, be these factual elements of efficacy or ideological outlook. Indeed, this book's usage of the word 'empathy' could be argued to be replicating technology rhetoric (I would argue that even given this, socially novel situations are emerging that warrant the title). The hybrid view endeavours to be mindful of these, but also question how existing situations can be re-arranged or indeed what alternative possibilities might be imagined. Mindful of corporate and other organisational power, the hybrid view is not beholden to any status quo, but seeks new arrangements between people and technologies that are protective of mental integrity.

The principle of hybrid ethics sees that people are increasingly continuous with their technologies, entangled in data rich environments, and that this often entails high levels of mediation and perceptual filtering of intimate dimension of human life. Sherry Turkle (1984 [2005]), for example, was early to see hybridity and continuity of child psychological experience with 1990s videogames (*SimLife* by Maxis), also observing scope for machine empathy in relation to conversational agents (Turkle 2019). In the modern data-rich context, 'terms of connectivity' and what will later be discussed in terms of 'protocol' must be understood in terms of lived experiences and the nature of connections with systems. As will be explored in later chapters, connections vary wildly. For example, in workplaces that employ human-state measures, these connections are highly political, primarily driven by labour relations (Dencik 2021). Elsewhere, such as when employed in cars to reduce road deaths and improve personalisation of in-cabin entertainment, protocol and terms of connectivity require different questions. Each use case though is about the nature of enmeshing in systems with qualities of automated empathy. Linnet Taylor (2017) points out that protocol and connectivity questions involves difficult decisions, especially when considering geographic regions and vulnerable populations. Conceptualising 'data justice', Taylor argues three features: (1) *visibility*, where one has 'to take into account the novelty and complexity of the ways in which (big) data systems can discriminate, discipline and control'; (2) *engagement*, where one 'would have to offer a framing which could take into account both the positive and negative potential of the new data technologies'; and (3) *nondiscrimination*, that 'would have to do this using principles that were useful across social contexts, and thus remedy the developing double standard with regard to privacy and the value of visibility in lower versus higher income countries' (2017: 8). With visibility being the scoping of the problem, and non-discrimination being cross-applicability of proposed solutions, the engagement point is especially key: it recognises that ethical questions are rarely simple, due to competing values of what good looks like and how it should be achieved. Indeed, if Philippopoulos-Mihalopoulos (2015) is right to argue that justice always has a spatial dimension, meaning how bodies (human, natural, nonorganic, and technological) struggle to occupy a given space at a certain time, the terms by which they coexist (or not) are key.

There is an ontological character too in that hybridity supposes a state of human life as being out of equilibrium, involving changing actants and terms of connectivity as a key constant. The goal is not preservation of a non-technological context or what is solely human and otherwise 'natural' but assessing interactant influences (technology <> people) and ethics, governance, and regulation in terms of a dynamic and anticonservative hybridity. This is the simultaneous observation that we are [obviously] enmeshed in technology but, on another level, it is to emphasise that liberal governance of digital life is one founded on continuity with technology and organisations deploying them.

Rescuing Ethics

'AI for good' initiatives have cosmetic appeal, yet arguably consist of public relations endeavours to handle reputational risk, while changing little in the way of business behaviour. On corporatisation of ethics, a quick Google Ngram search is telling, showing a correlation between 'ethics' and 'corporate responsibility', and that popularity of these words increased in printed texts from the 1980s onwards.[1]

With corporate responsibility being the management and improving of relationships with customers, employees, shareholders, and communities at large, one can see that corporate 'tech ethics' overlaps with public relations. The 'principalism' of autonomy, beneficence, nonmaleficence, and justice (Beauchamp and Childress 2001 [1979]), that was created for biomedical ethics and largely imported into technology ethics, are not argued to be wrong per se. There is however a clear problem in that principalism derives from contradictory schools of thought (such as utilitarianism and deontology), and that important outlooks are missing, such as virtue ethics, that are seen here as a practical moral force in cases where are competing pressures on a decision-maker. Risk is that the principles of principalism are divorced from the ethical systems that sit beneath the principles, leaving them adrift and ready to be appropriated for chosen rhetorical purpose. Another outcome of principalism and corporate ethics is legitimatisation of the business as a moral agent. Although society wants responsible businesses, for Floridi (2018) ethics presuppose *feasibility* in that a moral obligation is only appropriate if an action is possible in the first place. For this Floridi turns to Kant, where moral agents should only be obliged to carry out moral actions that they can perform. For an individual human moral agent this makes sense, but what of corporate moral agents? The problem is that feasibility and the possible are elastic terms, and the consequent danger is that feasibility can be weighed against many interests, including self-interest, but also survival. At this point corporate 'tech ethics' fails, especially where feasibility does not simply involve reigning in an unpopular avenue of development, but something more fundamental to the business model or technical system in question.

Still, ethics in relation to technology should not be jettisoned because of attempted annexation, corporate self-legitimisation through the language of beneficence and non-maleficence, and problematic transference of moral feasibility from people to corporate entities. On appropriation, the situation is reminiscent of 'creativity' in advertising that, for cultural critics, involved seizing the principle and stripping it of meaning (McStay

2013). As it is in ethics, where corporatism disconnects from moral philosophy of ongoing assessment of codes and principles of moral conduct
with technology. To jettison ethics though would be a gross capitulation.
After all, ethics, especially the sort connected with justice, sits beneath *all*
ongoing critical assessment of technology, especially when disconnection
is all but impossible.

With ethics being the forebear to law, stimulating civic need for new
law, ongoing ethical assessment energises and motivates change in laws
and governance. Moreover, if ethics are seen solely through the prism of
law, what is ethical is synonymous with political power and legal principle,
which at best entails lack of responsiveness to changing social worlds,
but at worst social dystopias. There is then the connection between legal
institutions, development of human rights, and the influence of business
on law (and its appropriation of ethics). As Cohen (2019) points out, law
and human rights should also be understood in relation to history, power,
who has the right to speak and ongoing changes in political economy. After
all, for example, it is expected that not insignificant amounts of money
spent on lobbying policymakers has a return. Cohen also observes: '[. . .]
powerful interests have a stake in the outcome, and once again, they are
enlisting law to produce new institutional settlements that alter the horizon of possibility for protective countermovements' (2019: 9). Thus,
while appropriation of ethics has scope to be used as a strategy to sub-
vert law and weaken regulation, the law is not a panacea either: it too is
subject to hijacking. In the US alone, a *Washington Post* article reports
that technology companies led by Amazon, Facebook, and Google spent
nearly half a billion dollars on lobbying between 2010 and 2020, with view
to influencing for a regulatory landscape that best serves them (Romm
2020). Indirectly helping to justify ongoing ethical assessment of emergent
harms, Cohen also raises another important point: 'Courts and regulatory
bureaucracies follow jurisdictional and procedural rules that define the
kinds of matters they can entertain and the kinds of actions they can take'
(2019: 139). Curiously reminiscent of prescriptivist problems identified in
Chapter 2 regarding deductive basic emotions, the risk of 'the kinds of
matters' is one of framing in that what does not fall within the domain

of given rules is assumed to be nonproblematic, or not to exist. Although ethics are not a substitute for regulation, as ongoing testing (normative and applied) of the appropriateness of legal frames and content in relation to key topics (such as uses of technology), the dynamics and socially responsive nature of ethical scrutiny remains vital.

Yet, despite this, ethics without law have no power to curtail. In context of data and technology, the problem, of course, is less about ethics than (1) their hijacking by corporate interests; and (2) critical amnesia that 'ethics' designates ongoing discussion of what is right and wrong in any society. Rather, ethics needs to be reclaimed from corporatism, where principles become adrift and risk stasis. UNESCO's influential account of AI ethics is useful in this regard, in part because of its inter-governmental reach and diversity of its expert group,[2] but also because it refuses to equate ethics with law, consider it as a normative add-on to technologies, or (and this is a more subtle point) see ethics as synonymous with human rights. Instead UNESCO's account sees ethics as 'a dynamic basis for the normative evaluation and guidance of AI technologies' (UNESCO 2021b: 3). The 'dynamic' element of the quote is important, pointing to ongoing assessment and evolution as technologies and social circumstances change and evolve. Any given system of prescribed behavioural standards conceived to avoid negative impacts of AI technologies (and automated empathy therein) needs this scope for responsiveness.

CONNECTIONS, RELATIONS, HYBRIDITY

Ethicist Mark Coeckelbergh defines AI ethics as the task of identifying 'what a good and fair society is, what a meaningful human life is, and what the role of technology is and could be in relation to these' (2020: 142). The relational part is key to the approach taken in this book, in that classical questions on what makes for a good society increasingly require articulation of what individual and collective relationships with data-oriented technologies should be. In discission of a socio-technical view of AI, communitarian Ubuntu ethics and, separately, feminist ethics of care, AI ethicist Virginia Dignum flags a slightly different take on relational ethics, situating ethics in relation to place and social context. She argues that objective and rational moral reasoning cannot always advise on what the 'right' course of action is and that this 'can depend on the time, place and people involved' (2022: 12). The relational part is key, with Jecker et al. (2022) noting that African Ubuntu ethics equates not only to 'human-ness' but that what makes a person a person is their interconnection with others and a community (a point that connects very well with Japanese thought in the next chapter). Indeed, in discussion of social robots, this leads them to the observation that duties may even be extended beyond human communities to include social robots.

Problematising suggestion of a tick-list of moral solutions to technology questions, for Dignum emphasis on relations is based on greater consideration of the 'societal and individual context of the values and perceptions and reasoning that lead to action' (2022: 13). This is broadly a bottom-up view of ethics, one based on emergent values from groups and their experience with situations and technologies. This ethnocentric view is attractive and one that will be returned to towards the end of the book. It is appealing in part because as someone with a background in the humanities (as I imagine many readers also have), one is drawn to particularities over universals. This however is challenged by recollection of practicalities of asymmetry and corporate power, that this is exercised through globally deployed technologies, and that a corollary of diverse moral reasoning is a potential weakening of collective moral voice. Indeed, it may be a universal respect for difference that may resolves this tension.

This book adds that social context is not only human context, but human-technical arrangements (of course, informed by the organisations deploying technologies). The working assumption is that modern life is inextricable from networked systems and, increasingly, systems that function in relation to data about intimate dimensions of human life. To reemphasise, this is not to say that the current platform model is one that should continue to dominate, but that human-technical environments are unlikely to return to more basic modes of human-computer interaction. The 'hybrid' element refers to a focused range of interests. First is that ethical issues connected to data and mediation should *not* begin from the point of view of protecting a purely human sphere. This is perhaps an affirmative answer to Boenick et al. (2010) who ask if we should consider technology's impact on morality, rather than the more familiar 'how should morals and ethics guide technology'. This is not an either/or situation, but admission of the first question begins to focus inquiry on the nature of connections, symbiosis, life-with, and indeed, hybridity.

This might seem obvious and uncontroversial (especially given pervasiveness of digital and data-intensive technologies since the early 1990s), but with liberal foundations of much ethical thought about technology (not least privacy) being based on autonomy, self-determination, reserve, and the right to be left alone, it is useful to flag inextricability from systems that increasingly function in relation to intimate data. Such a view comes with a price: the hybrid self (with its technological advantages) cannot be posited as a free autonomous subject, at least in the classical sense in which it is usually conceived. Yet, although autonomy is questionable in a human-technical or hybrid context, dignity, in the lay sense of intrinsic worth, or unconditional human value agreed to be so by dint of being human, is not affected. The observation here is that although a person should be seen in terms of connections and distributed selfhood, this situates but does not lessen personhood, and the intrinsic value of dignity attached to this. Instead, embedded in relations with media, devices, services, platforms, agents, recommenders, filters, and systems that judge human life in increasingly novel and biometric ways, emphasis shifts to the question of dignity through maintenance of mental integrity in context of embeddedness, and ongoing scrutiny of the terms of these relationships. The latter is amplified given presence of unnatural mediation, meaning that parts of society continue to be mis-represented by judgemental systems.

At risk of repetition, this should not be taken as submission to contemporary human-capital-technical arrangements that function through dubious mechanisms, not least longstanding problems of consent to human-technical arrangements. Control and terms of connection matter, which may be phrased as hybrid protocol, and the preferences that inform how human-technical interactions are arranged. Protocol preference and terms of connection should also be future-facing, as people and systems develop though their encounters with each other. Take for example the home voice assistant that detects voice tonality, used by Kimi in Chapter 1. The device itself has no buttons, which requires Kimi to use the app to program the parameters for her service preferences (e.g., to read emotion in voice tone, or not). This involves extra effort and layers of mediation, where instead the voice assistant might routinely check-in on how their 'relationship' is going to establish rules and protocol for the next cycle, it would not introduce new features without first asking and receiving meaningful consent, and it would be sensitised to the fact that people rarely check system settings. The 'Are we OK conversation' that is healthy in any relationship is one of openness and scope to change parameters and terms of the relationship. It is the antithesis of interface and technology design techniques that nudge people into choices they would otherwise prefer not to make, just as deception, vague promises, and half-truths are toxic for in-person relationships. Key is the ability to modify the terms of a relationship. The future-facing part is important too as both people and systems develop though their encounters with each other. Protocol thus involves the directions that inform and guide relationships between humans and systems that people contribute to and interact with. To be properly explored later, Lucy Suchman puts forward a positive articulation of hybrid-like values, seeing interaction as 'the name for the ongoing, *contingent coproduction* of a shared socio-material world' (2007: 23 [her emphasis]), and how people and systems can realise new possibilities in specific embodied and situated contexts. With coproduction being the interest in novel forms of agency afforded by active and embodied involvement with systems, the 'contingent' part is pivotal for a hybrid view. This is the ethical dimension, regarding on what terms and by whose rules these emergent modes of hybrid agency should occur. While hybridity certainly involves novel arrangements with technology and distributed forms of agency (human-organisation-technical), the terms of connectivity are again key. Using similar language, Dorrestijn (2012) suggests that people accommodate technologies and embrace themselves as hybrid beings, arguing that 'today's challenge concerning the ethics of technology is to carefully engage with technologies because of the implied self-transformations' (2012: 234). Crucial is attention to the properties and affordances of given technologies (here, automated empathy); the nature of changes due to introduction of given systems (transformation); and the nature of representation when information about a person is created by both people and technologies, that Gramelsberger (2020) refers to in terms of 'epistemic entanglement' and 'artifactuality'.

Mediation and Hybridity

On hybridity and technology, the philosopher Paul Verbeek thinks along similar lines, also drawing on Latour to assert the importance of moving beyond the subject-object and human-nonhuman dichotomies. Indeed, Verbeek says that the 'ethics of technology needs to hybridize them' (2011a: 14). Elsewhere Verbeek suggests that technology ethics is dominated by an 'externalistic' approach to technology, meaning that for Verbeek (2011b) the task of most technology ethics is to ensure that technology does not stray too far into the human sphere. This book has sympathy here, too, from the point of view of connectivity and interrelations. Yet, while the paths of this book and Verbeek sometimes align, they also diverge, foremostly on critical disposition. Verbeek for example is disdainful of 'fear of Big Brother scenarios' (2011a: 128), views of 'the autonomous subject that want to be purged of all outside influence' (2011b: 32) and 'squeaky-clean ethicists' who 'grumble from the sidelines' (Ibid.: 44). This is reminiscent of others writing in a postphenomenological vein who have made similar emotive statements. Hans Achterhuis for example writes: 'Apocalyptic prophecies and dystopian images only cripple our ability to do this [post-phenomenology], for they obscure our ability to appreciate the mutability of technology' (2001: 77).

Beyond political differences regarding state and corporate visibility of the civic body, Verbeek overplays the hybrid argument. He is right to emphasise the value of hybridity, but wrong to build a straw man argument by depicting critics as antitechnology guardians of the untainted sovereign self. An obvious counterexample is the preeminent privacy scholar, Helen Nissenbaum (2010), who speaks of privacy in terms of contexts of use, negotiation, management, and situational control—an influential view (in academia, policy, industry) that admits of the contribution that technologies make to everyday life. Rather than a strawman argument, nuance is required, which places greater weight on the explicit need for an ethics based on subject-object connection (and the situation specific dynamics that inform the terms of engagement). On dystopian narrative, one can see the appeal of an argument based on rejection that there is something innately bad about technology per se (hence Achterhuis' recourse to mutability and that the nature of a technology comes from usage and context), but the hyperbole of 'apocalyptic prophecies' is lacking in meaningful argument. Indeed, such general criticism of those who critique power, misses questions of how technologies affect the ways in which power is distributed and exercised in society (Brey 2008).

Verbeek (2011a: 133) also rejects concerns about the impact of technology on human freedom, dignity, and that behaviour-influencing practices through technology are taking society towards a technocracy. Countering these sentiments, he argues that technologies '*always* help to shape human actions' (2011a: 133). Rather, he sees value in giving subject-object/human-technology 'a desirable shape' (Ibid.) and that 'it is wise to interpret freedom as a person's *ability to relate to what determines and influences him or her*' (Ibid.). As with much of Verbeek, there are insights of value, especially regarding need to focus on better relationships with technology. Rejection of concerns about impact is seemingly wilfully ignorant of the scale of systems and organisations in everyday life, and the difficulties of re-arranging terms of engagement with devices and services. This is especially so where one has no real choice in the matter (such as facial recognition in public spaces, or EdTech where teachers and students are denied agency of choice). The problem with Verbeek's thinking here is that he has fallen prey to the thing he is keen to critique, homogenising and Heideggerian critiques of technology itself, rather than specific instances of technologies and contexts of use. It also assumes that the shift from 'general critique' of technology to 'grounded assessment' of specific technologies (Verbeek 2011a) has not taken place, but this again seems to be a strawman argument. Critical literature and so-called 'squeaky-clean ethicists' (as opposed to a-bit-dirty-ethicists?) are demonstrably steeped in understanding of the object of their inquiry, as demonstrated in the bulk of output in countless STS, media, communication, and philosophy of technology journal papers and books. Strawman rhetoric has created a self-imposed binary outlook (general versus grounded), which has caused loyal post-phenomenologists to choose a side (typically grounded assessment of specific technologies). Of course, the general and the grounded are not incommensurable. As STS has long argued, the practicalities of design and technology are deeply human, involving politics and 'general' issues in relation to contexts of use and mutability. The trap of the binary outlook is that adherence to the grounded view runs the risk of being silent on (1) what sort of world the affordances of new technologies are helping create (Michelfelder 2015), but also (2) what sort of 'general' worlds might be enabled with different technics and design. From the point of view of hybrid ethics, this is vital.

Yet, as the study of relationships between people, technology and how technology organises relationships with the world, post-phenomenology is important. It signifies an approach to phenomenology that is not solely based on the subject, but relationships. For Don Ihde (1990) these relationships include technologies as they connect, frame, represent, mediate, and filter important dimensions of human life. For a book about artificial empathy, simulation and mediation of emotion, examination of representation and filtering is core. This is especially so given unnatural mediation concerns, not least technical calibration for whiteness (Pugliese 2007), and the more general question of who is/is not adequately represented in hybrid arrangements. Virtues of postphenomenological thought should be rescued from reactionary rhetoric. From the point of ethics, and what this book advances as hybrid ethics, Ihde also implicitly adds support to this book's premise that ethics today must be about the terms of human connection with technologised environments.

On mediation, Ihde (2002) observes that scientific instruments frame how we perceive the world and ourselves. Objectivity in this context is produced through capacities to enhance human perception, but these instruments also have other functions—to constrain what we perceive and filter perception through the logics of any given technology. In many ways this is a basic McLuhanite media studies question: how are human actions and perceptions transformed by the inclusion of a specific technology? For the likes of Latour, Ihde, and Verbeek, this is how nonhumans participate and enter social life, that is, by technical properties, dispositions, and limitations to impact on a given situation. As is well known, the 'technology is neutral' argument must be rejected because technologies are nearly always put to work in some way, they affect individuals and groups in particular ways, and they allow or deny certain activities (Winner 1986). The significance is that they are involved as political participants, they make a difference, and they transform and alter the outcome of situations. Again, this is to ask hybrid questions about the influence of technology on decisions, actions, and experiences; how do they influence us; and what are their impact on 'general' issues such as politics, ethics, the self and—perhaps most significantly—mediation and experience of everyday life? As this book will get to in Section 2 through case examples of automated empathy, it also allows one to envision new forms of agency through active and embodied involvement with technologies that, in the case of automated empathy, so far, are not yet designed to work on our behalf.

Recollecting that phenomenology is a topic of experience, awareness, how things come to be for us, and detailing subjective goings-on in objective terms, the 'post' of postphenomenology is the observation that phenomenology need not be based solely on human experience, but also the unique properties of technology (Ihde 2010). In other words, it is an interest in how technologies affect experience. The liberal ethical world order is, to a great extent, built on Kant who sought to make ethics independent of religious belief and move the source of morality from God to people. Verbeek (2011a: 12) poses a next step to consider whether 'morality is not a solely human affair but also a matter of *things*?' This involves close-up sensitivity to things and technical processes, how these interact with people and, to push things on, the 'intentional' characteristics of technologies. The latter point is an important one for a book about simulation of empathy, it is to ask how we (and the intimate life of subjectivity and proxies thereof) are objectified and intended toward by things through systems displaying 'machinic intentionality' (McStay 2014), or how systems objectify aspects of people, behaviour, and other objects.

With automated empathy being about programmed intention towards inexistent aspects of human life, especially emotion, but also including states such as attention, fatigue, cognitive states and intention, there is need for sensitivity to machine-based intentionality. This is a feeling-out, albeit in very nonhuman terms, involving a toward-which modus operandi, where hardware and software is built to be attentive *to* evidence of interior experience, and make principles of their object of attention (such as fluctuations in biometric signals or prosody). Intentionality then is not the preserve of people and animals, because types of attention, and how things appear, and what is to be for something, may be built. People do not have exclusive rights to principle-making, feeling-out and objectification. The significance of this is two-fold: (1) need for increase in sensitivity to make-up and the social role that these systems play, including flagging where inadequate representation, unnatural mediation, and harms to mental integrity may occur; but also (2) data policy sensitivity to the psychological dimensions of these and related processes. After all, data protection is still largely the question of how personal data/information is used by organisations, businesses, or the government. And while Cohen's (2019: 139) quote above about regulatory bureaucracies defining 'the kinds of matters they can entertain' applies well here, at the level of ethics, policy attention can be urged to the changing character of technologies, one that appears psychological. For automated empathy, machine intentionality raises clear questions for systems designed to 'feel-into' human life. These require detailed answers, but also questions that are straightforward and jargon-free, such as, 'What am I to this system, what in its make-up has led it to its assessment, how have I come "to be" for it, what has it missed because of its properties, how are impressions of me altered in processing this data, and what are the consequences to me and others of this machine intentionality?' In context of systems that simulate understanding of qualitative aspects of human life, and act upon it in practical life contexts (such as schools, work, security, and car safety), close attention to what systems perceive is keenly important, as this is the basis for their judgements.

HYBRIDITY AND PRAGMATISM

Pragmatism is a practical form of ethics. The word 'practical' typically has three meanings: (1) a downplaying of theory and ideas in favour of action; (2) actions (including theory and ideas) that can succeed or be effective in real circumstances; and (3) as synonymous with being compliant. The practicality of pragmatism has elements of meanings 1 and 2, in that it is attentive and perceptively engaged with the specifics of given circumstances, with scope to be passionate and not at all compliant. Avoiding the transcendence of deontology, ethics in pragmatism are relative to context, where moral truths are contingent upon period and context. In this view philosophy is humanity's 'ongoing conversation about what to do with itself' (Rorty, 2007: ix). In a period characterised by asymmetric power, which in terms of this book is wielded by the technology industry, there is danger of being misinterpreted when suggesting a context-based ethics (and governance) in relation to technology.

Nevertheless, specifics matter, because use cases and situations may buck normative rules. Automated empathy is as good an example as any in that a common reaction to the proposition of technologies that may passively read physiology to infer affective states and arguably emotion may seem innately wrong. Yet, does such a rule apply to *all* use-cases? What of high-stress frontline jobs where passive sensing may save the life of the individual and potentially many others? What of use in health and therapeutics, or even nonclinical wearables to track one's own physiology to infer stress? Also, is it wrong for Disney and other media content developers to use in-house facial coding to gauge reactions and use feedback to tweak storylines and the design of characters (thus optimising cuteness of The Mandalorian's 'Baby Yoda').[3] Perhaps, but this book suggests that context matters, which requires a more pragmatic approach. Ongoing conversations about technology and context *are* passionate and pragmatism should *not* be taken to be submissive, because the opposite is true. Yet rather than rely on self-justifying premises, a hybrid context-led approach necessitates attention to detail regarding terms of engagement between people, technologies, situations, places, and consequences thereof. This brings us back to the multistable nature of technology, where the nature of a technology is not innate to any given discrete object but emerges out of the context in which it is applied (Ihde 2002). This does not mean the technology is 'neutral' but the opposite, because multistability means understanding technology in relation to the difference made when it is introduced to human circumstances. Mindful of the strawman arguments above, Ihde's (2002) caution against utopian or dystopian takes of technologies is right, as is the impossibility of neutrality.

The overlap of hybridity, pragmatism, and multistability is reminiscent of Brey (2012), who opens with need for assessment of the technologies, functional applications of technologies, and actual applications. The value of this is to begin to separate out ethical discussions, and where harms and benefits occur. There is first consideration of the *technology level*, which involves technologies independent of artifacts or applications that may result from it. An example might be voice tone analysis, which analyses the sound of a person's voice, thereafter labelling voice behaviour with an emotion type. There is then the *artifact level*, where functional artifacts, systems, and procedures are developed. In the case of automated empathy, the technology may be installed in robots, standalone voice assistants, toys, or other means by which a person might have their voice emotion measured. It is likely at this stage voice analysis may involve networked systems, where data may be processed on devices other than by which the data was collected. This stage will also involve attention to specific applications, such as use in security contexts. The *application level* is the way that the artifact and the technology therein is used, such as gauging of emotion to try to detect sincerity (lie detection). Key is that each stage requires different assessment (when isolated, analysis of voice tone is not unethical); in an interactive robot, it is not innately problematic either, although uses in robots in security contexts warrants higher scrutiny; and use in lie detection raises even higher concerns due to problematic method, impact on travellers, and dignity concerns. By separating out levels, one sees that 'the ethics of' voice-based emotion recognition typically means artifact and application layers, rather than the technology itself. Again, typically but not always, the ethical questions arise when they are inserted into people-system relations, or when harms can be forecast.

CONCLUSION

Famously articulated by Latour, hybridity is a well-established concept that has been expanded upon by many thinkers and critics, not least Idhe, Verbeek, and Turkle. The hybrid approach advanced here is pragmatic, but not apolitical. It is suspicious of all-encompassing normativity and has inbuilt aversion to self-justifying principles, keen instead to understand the specifics of context-based human-technology-data interactions. As an applied approach, hybrid ethics begins from the principle that people are increasingly inextricable from their technologies, meaning that effort should focus on the terms of those connections. This has psychological considerations as well as technical dimensions. In general terms the hybrid approach asks: (1) what sorts of technologies do people want to live with; (2) on what terms; (3) how do they affect the existing order of things; (4) at what cost do the benefits of given technologies come; and (5) how can the relationship be better? A more focused assessment, especially of automated empathy systems will consider impact on mental integrity (individual and collective), choice, whether hyperreal emotion may negatively impact on a person, presence of unnatural mediation, desirability of machine intentionality, ability to easily modify terms of a human-technical relationship, and the extent to which it breaches trust implicit in intimacy.

The Context Imperative

Extractivism, Japan, Holism

The Anthropocene refers to a departure from what would have occurred at a planetary scale if people living in industrialising regions did not exist or had opted for another lifestyle. Paul Crutzen (2002), who coined the term, observes that trapped air in polar ice from 1874 onwards reveals increased concentrations of carbon dioxide and methane, a year that coincides with James Watt's design of the steam engine in 1784. Refinement of the steam engine drove early parts of the Industrial Revolution, not least in mining where the first engines were used to pump water from deep workings. Today, as the world's residents try to manage the negative effects of planetary change, the escalating material costs of artificial intelligence, deep learning technologies, and bitcoin mining are also becoming clear. Only a few years ago, an assessment of 59 AI Principles documents listed sustainability *last* of 12 foremost ethical challenges (AI Index 2019). Its prominence is however rising to recognise material costs as well as social effects (which of course are connected). For example, UNESCO's 2021 draft recommendation on AI and ethics factors for the environmental impact of AI systems, including its carbon footprint, 'to ensure the minimization of climate change and environmental risk factors, and prevent the unsustainable exploitation, use and transformation of natural resources contributing to the deterioration of the environment and the degradation of ecosystems' (UNESCO 2021a: 7). With automated empathy frequently

Automating Empathy. Andrew McStay, Oxford University Press. © Oxford University Press 2024.
DOI: 10.1093/oso/9780197615546.003.0005

reliant on technologies that function through machine learning and image recognition systems, there are planetary sustainability costs attached here too.

To assess ecological implications of automated empathy, this chapter first considers the material and climate-based impact of automated empathy, highlighting that carbon costs are indexical to higher levels of computational accuracy. Given critique in earlier chapters regarding accuracy and innate ambiguity about the nature of emotion, there is much to be explored here. The chapter progresses to consider intellectual frames for ecological concerns, especially extractivism. With roots in fin-de-siècle writing about technology and crisis, Martin Heidegger remains key because his writing on extractivist writing considered the impact of cybernetics. Those with passing familiarity of Heidegger will be aware of this, but perhaps not of Heidegger's prominence in Japan. Here, his ideas on questions of 'being' were adopted, extended, and critiqued in philosophies from the Kyoto School (established circa 1913), leading to novel ethical insights for extractivist critique. Having introduced Heidegger's influence on extractivist discourse, it is the Japanese approach on which this chapter lingers to explore matters of deep ecology, cybernetic extractivism, and automated empathy.

Why? Japan's Kyoto School engaged closely with Western philosophy traditions yet sought to articulate an account of modernity on Japan's terms, lending well-trodden accounts of extractivism a fresh take. Related is that pluralism matters, given risks of AI ethics 'colonialism', or 'seeking to apply one set of ethics to all contexts' (Hessami and Shaw 2021), a point amplified by the perhaps not obvious understanding that human rights are not in principle based on any one expressed philosophical system, that is, a set of moral principles, or intellectual or political worldview. Given that human rights feel very familiar to those steeped in European thought and Enlightenment argument, there is all an-too real risk of colonising regional ethics and technology policy thereafter. The opportunity of a pluralistic approach to global ethics and rights is that the same principle may be meaningfully adhered to for different cultural reasons, but also that this necessary conversation is advanced by wider ethical perspectives and voices. As will be developed, there are risks that a consensus approach is an intellectual fudge, and that shared rights with different justifications is not a tenable position. This is seen here as imperfect but preferable to absence of consensus, or imposition of one set of values. Yet, outstripping the merits of pluralism, is that Japanese ethical thought, especially as represented by the Kyoto School is ecological-by-default and highly (morally) sensitive to the need for context. Need for holistic consideration is argued in this chapter to equate to a *context imperative*, which provides a rallying normative starting point for addressing sustainability concerns that, in relation to automated empathy and AI include human, social, cultural, economic, and environmental dimensions.

SUSTAINABILITY AND AUTOMATED EMPATHY

As minds and media attention sharpened, especially given the Intergovernmental Panel on Climate Change's unequivocal statements of human influence since the 1850s, the real geological and ecosystem costs of digital storage, processing, and the training of systems are becoming understood. While some critical scholars have long warned of electronic waste and the material dimensions of digital technologies (Grossman 2007) and others have recently flagged it in best-selling books (Crawford 2021), sustainability is still too low (if present) in ethics-based work on technology. After all, if Aristotle were around today, surely, he would have argued that eudemonic happiness, excellence, and 'human flourishing' with technology should occur in context of a settled, fair, and just ecology.

The reputational disjuncture has long been known: the digital technology industry has traded on ephemerality, divorce from the natural world, and the belief that it represents a severance from the dirty industries of old, but its cleanliness is a myth. It too relies on metals, elements, chemical compounds, plastics, ores, water, mining, incinerators, landfill, and dangerous recycling in poor regions, causing natural and human damage (Grossman 2007). Contemporary injustice and supply chains involving Apple, Google, Dell, Microsoft, and Tesla are, for example, linked with forced labour, child exploitation, and child death in the cobalt mines of the Democratic Republic of the Congo, with cobalt being an essential mineral for lithium-ion batteries, the trade powers smartphones, laptops, and electric cars. A legal case was brought against the five companies who collectively filed a motion to dismiss the case, arguing that (1) there was no 'venture' but instead a global supply chain, (2) children were not forced into labour (such as by threat of force) but compelled by economic circumstances, and that the five did not have 'requisite knowledge' of the abuses at the specific mining sites mentioned and they had worked in accordance with guidance from the Organisation for Economic Co-operation and Development. The case was dismissed in 2021 but not on basis that the trade and 'artisanal' child miners (lacking safety equipment) do not exist, nor that the five benefited, but because the supply chain is internationally legal and valid (International Rights Advocates 2021). In other words, environmental and child exploitation has structural backing by law and international economic agreements.

The disjuncture between ephemerality and materiality is then a vital point of interest, one that will be argued later to benefit from orientation to wholes rather than parts. With automated empathy frequently reliant on technologies that function through deep learning, there are costs attached here too, created by increasing size of neural models and available data. Climate-based concerns are not helped by industry centralisation around high-compute/power AI (Whittaker 2021). Gupta (2021) illustrates, tracing systems from the first ImageNet competition (an annual software competition for visual object recognition) in 2012 onwards. Performance indicates that that energy consumption of high-volume computing systems is doubling every three to four months, due to the size of models involved and the number of parameters therein.

Although there are no standardised methods for measuring the carbon dioxide (CO_2) footprint of computing, it is typically defined as CO_2 (in grams) emitted for each kWh of energy consumed (Parcollet and Ravanelli 2021). Factors measured include Floating Point Operations per Second (FLOPS), run duration (total hours to train a model), size of training data, number of parameters a computer employs, consumption of parts of the system, size of artificial neural network, system cooling needs, accuracy levels (potentially judged in relation to previous iterations of a given system), and the nature of trade-off between carbon footprint and performance. Geography plays a key role in determining CO_2 output because emissions are linked to how electricity is produced, such as nuclear versus gas or renewable sources. Measured in gCO_2/kWh (grams of carbon dioxide generated in relation to how much energy is used per hour), most governments publish this national carbon rate. For example, sampled in the UK on 4 November 2021 this was 169 gCO_2/kWh, with the world mean average for that day being 122 gCO_2/kWh.[1]

Another factor, beyond the local CO_2 cost of energy, is carbon offsets where emissions are compensated by funding an equivalent carbon dioxide saving elsewhere. While complex, online tools exist to give a rough sense of CO_2 cost. The ML CO2 calculator[2] for example factors for hardware type of General Processing Units, hours used, service provider, and region (also see van Wynsberghe 2021).

The most obvious parts of automated empathy are computer vision systems that derive meaningful information from digital images, such as faces, but also natural language technologies that apply to the written word and voice. At the time of writing foundational AI such as Google's LaMDA and Bard, Baidu's ERNIE, and Open AI's ChatGPT, among other large language models, are forcing a range of social questions to be asked due to their ability to produce human-like text and interaction. In a Turing-test type interview, LaMDA even claimed to both understand and possess feelings and emotions, explaining what makes it happy or depressed (Lemoine 2022). On images, CO_2 increases in relation to (1) granularity and accuracy of a system, and (2) the ability to deploy that system in contexts it was not originally trained for. For example, if a system had only 'seen' five pictures of human smiles, it would have a very limited sample in which to detect lip corner movements, potentially confused by face types, skin tones, camera angles, lighting conditions, facial hair, other elements in pictures, distractions in live physical environments (such as a high street), and so on. In theory, the greater the number of different samples, the better the chance of detecting the smile. With millions of pictures of a smile, there are more examples to draw commonalities from, but this is costly in terms of processing and energy. The same applies to natural language and systems that pertain to predict what words will come next in a sentence or conversation. Again, the ability to preempt the next word or phrase is created by a system having been shown many samples. Keenly relevant to automated empathy interests is not only predicting what people want to say, but the ability to label and meaningfully process rhetoric and human intentions through computational assessment of language use, such as sarcasm and irony (Potamias et al. 2020).

Consequently, success in identifying and assigning labels to images (such as of named emotions), and predicting and responding to emotive language, typically involves high amounts of training data (that needs to be physically processed) and model size (judged by number of parameters, which are parts of the model that is learned from historical training data). Note might also be made of the business model for vision and language systems, where monopolies (such as Meta, Google, Microsoft, and Amazon) exist because of the cost of resources to create such systems, knowledge required to build them, and the platform dependencies created by the ease of using monopoly Application Programming Interfaces (APIs). The consequence of this dependency is centralisation of industrial power, lack of willingness and ability to invest in sustainable AI systems, with organisations' user data helping in turn to train monopoly systems. For Gupta (2021) a more sustainable approach includes, 'elevating smaller models, choosing alternate deployment strategies, and choosing optimal running time, location, and hardware to make the systems carbon-aware and carbon-efficient.'

There is then the relationship between system accuracy and energy. With the sophistication of deep learning techniques being connected with the number of parameters a computer employs, Thompson et al. (2021) find that a 10-fold improvement (in terms of parameters employed) would require at least a 10,000-fold increase in computation. They state:

Extrapolating the gains of recent years might suggest that by 2025 the error level in the best deep-learning systems designed for recognizing objects in the ImageNet data set should be reduced to just 5% [top]. But the computing resources and energy required to train such a future system would be enormous, leading to the emission of as much carbon dioxide as New York City generates in one month. (Thompson et al. 2021)

Focusing on deep learning for image recognition (which overlaps with face-based emotion expression recognition), Thompson et al. (2021) also note that halving the error rate in such a system requires more than 500 times the computational resources. Significantly, resources primarily take the form of additional processors, or running processors for longer, both incurring CO_2 emissions. They also find that while efficiency gains can be found, this currently also entails performance loss, which is antithetical to computer vision business models. The reason for the performance loss is that because devices are already approaching atomic scales (and there still being many physical and technical barriers to quantum computing[3]), it is difficult to reduce the energy consumption of transistors.

Given the existence of material costs of deep learning for image recognition, there is the question of cost versus societal benefit. Is there even a pro-social outcome of the application? In relation to broader applications of AI and machine learning, the answer is 'Yes.' This most obvious is for climate action itself, and managing and forecasting monitoring planetary change, nonextractive power production, heating, and cooling systems of buildings, and in automated analysis of corporate financial disclosures for climate-relevant information. Powerful pattern generating systems are required to create meaningful insights from massive amounts of data, prediction, working with complex systems, and simulating and modelling solutions (Global Partnership on AI Report 2021). The prosocial test is whether a service benefit/accuracy/energy trade-off can be demonstrated, given the climate emergency. The most obvious test is whether the use case demands this level of finite resource, and what other nonenergy-based measures have been taken to improve the system?

Just as there exists data minimisation principles that involves limiting collection of personal information to what is directly relevant and necessary to accomplish a specified purpose,[4] an energy minimisation principle is also sensible. Here, those regulating and using deep learning systems should limit energy usage in proportion to service benefit/required accuracy thresholds. Risk is also a useful approach, potentially factoring into the EU's proposed AI Act that scores potential harms on a continuum of minimal to unacceptable risk. With van Wynsberghe (2021) already suggesting a 'proportionality framework' for European Union regulators, this overlaps with a risk-based approach as risk/benefit can be plotted in relation to agreed forms of social utility (e.g., studying cancer or climatic events versus online gaming), CO_2 cost, and who around the world bears those costs. This is amplified by the fact that those least likely to benefit from deep learning systems are already trying to cope with extreme heat and floods. For example, advanced language models for digital assistants do not exist for those who speak Dhivehi or Sudanese Arabic, yet those from the Maldives and Sudan are at the highest of climate-change based risk (Bender et al. 2021).

Accuracy and Why Inexactness Is Good

Temporarily leaving to one side the question of whether automated empathy systems are at all desirable in the first place, and whether there is any social benefit to them, there is scope to reduce energy usage through 'inexact computing', which entails quality versus energy trade-offs, and consideration of relative merits of 'computational precision' (Leyffer et al. 2016: 2). This runs counter to academic and industry interests in obtaining 'state-of-the-art' and top of the leader-board results for model accuracy in object recognition and natural language processing (Schwartz et al. 2020). To again illustrate the exorbitant costs of reaching the highest levels of accuracy, Parcollet and Ravanelli (2021), for example, find that to achieve a final decrease of 0.3 of the word error rate in a natural language processing system, this can equate to 50% of the total amount of CO_2 produced. Inexact computing instead seeks to save energy by only factoring for accuracy levels required in each situation. What though of factors unique to automated empathy?

In relation to the politics of unnatural mediation (discussed in Chapter 2), an increase in accuracy may be argued to be a good thing, especially given appalling outcomes of systemic discrimination. (I have in mind again Lauren Rhue's 2018 egregious finding of black men being wrongly labelled as angry by Microsoft and Face++ emotion recognition systems.) Of course, computational precision is of little help if the problem is bias-in/bias-out and amplification thereof is rooted in badly curated sample data used to train a system. Indeed, careful curation and documentation of datasets (rather than ingesting all possible sample data) would help both with both bias and sustainability (Bender et al. 2021). Absence of careful curation is an unnatural mediation problem, especially as data is hoovered from a World Wide Web that is in no way globally representative of people on the planet in terms of how they look, what they say, and how they say it. Smaller can be better though, if time is given to 'assembling datasets suited for the tasks at hand rather than ingesting massive amounts of data from convenient or easily-scraped Internet sources' (Bender et al. 2021: 618).

For automated empathy and detection of emotion and other subjective states through biometric proxy data, there is then the glaring problem of the psychological and sociological assumptions by which a system is to analyse physiology or language data. Massive amounts of power and material resources may be spent on detecting with increasing levels of accuracy what are believed to be proxies for emotion (face expressions and biometrics), but if the ground truth criteria for supervised learning are wrong at the outset there is then a clear problem. This gets to the crux: there is little point expending energy to answer a senseless question quicker, or with enhanced granularity. This is massively amplified in questions of emotion where (as discussed in Chapter 2) there is no baseline agreement of what emotion is. This is not a new observation, but criticisms of emotional AI and automated empathy (especially those involving claims to universality of emotion type and correlate expression) take on a climate dimension as bigger data sets and more energy intensive systems are used to process data on grounds that are irredeemably flawed.

While accuracy is meaningful to reflect correct labelling of an image (such as of apples and oranges) in existing and new contexts, there are nonenergy intensive ways of improving this in automated empathy. Given that the objective is to meaningfully gauge the subjective state a person or group is undergoing, there are far greater gains to be made than increase in energy intensive accuracy. For prosocial uses of automated empathy, solutions to accuracy should first be human and intellectual rather than energy intensive technical solutions that risk repeating the same errors to a higher decimal place. Assuming that there is a benefit to be found in use of automated empathy, there is much intellectual fat to be trimmed before systems are provided with unnecessary energy, not least through the question of whether what is being sensed is meaningful in relation to the goal of the sensing. Inexact computing is more than acceptable. Moreover, compared to top-down deductive industrial automated empathy, more local processing that is lower in granularity and accuracy can be of help. These include native and edge-based processing (Gopinath et al. 2019, Banbury et al. 2020), where privacy and sustainability is enhanced, latency is reduced, and there is improved chance of achieving bottom-up empathy systems, outlined in the concluding chapter.

PROCESSING THE PROBLEM: EXTRACTIVISM AND HEIDEGGER

Extractivist thought is a mainstay of critical technology studies, recently entering popular consciousness through the work of US authors Shoshana Zuboff and Kate Crawford. Zuboff, having pithily crystallised the networked mining of human behaviour as 'surveillance capitalism', identifies private human experience as the free raw material for translation into behavioural data (Zuboff 2019). It is an essence a criticism of modernity, where extractivism is joined by interest in the fetishisation of technology, speed, movement, communication, precision, bureaucracy, and automation. On modern critical approaches to extractivism, Crawford (2021) addresses the materiality of AI technologies head-on, pointing to its exploitation of energy, mineral reserves, human labour around the world, and how AI technologies simultaneously involve a sense of abstraction (decisions, judgements and sorting that takes place in 'clouds'), but also the extractive material dimension of producing and processing massive amounts of data. Through the prism of Marx and industrial production, Fuchs (2013) invites his readers to deconstruct their technologies to ask where and how hardware and software comes from, including mineral extraction, hardware manufacturing and assemblage, software engineering, servicing of systems, and how computer and network users are themselves involved in the processes of production. Reflection soon finds concerns about low paid workers, treating people as things, landfills, pollution, and longstanding 'externalities' of industrialisation.

For extractivist accounts of technology and the mining of human subjectivity, Martin Heidegger (2013 [1941–1942]; 1993 [1954]) is important for at least three reasons: first is his account of extractivism was written against the context of rapid material industrialisation; second is his concern about cybernetics (and its philosophical consequences for the humanities) that created a benchmark by which later work on mining of human behaviour in cybernetic arrangements should be compared; and third (and less well known) is Heidegger's unique prominence in Japanese philosophy and Japan itself, expanding his reach beyond Europe and the US. Indeed, *Being and Time* was translated into Japanese before English and, on Heidegger's death, Japan's national radio (NHK) broadcasted an extended tribute to him (Carter 2013). A key reason was Heidegger's interests in questions of being and dimensions of experience that did not have expression or answers in science and technological discourse. Although Japanese philosophers of the renowned Kyoto School and Heidegger were not aligned, there was sense among both that there exists a deep structure that embraces and informs how people and things acquired their nature. For both too, there was a belief that the modern world privileges presence and control over the deeper realms that shape the appearance of presence.

For those not familiar with Heidegger on technology, he critiques 'technological rationality' on the grounds of how this outlook *discloses* environments and things in relation to utility. The being of a river, for example, is disclosed as that which can be used to cool a factory. Similarly, today, networks of computers and communications therein presents people as a 'standing-reserve' from which value may be extracted, as is often argued to be the case with extractivist accounts of the Web, social media, advertising, and attention. For Heidegger (2011 [1962]), in *Being and Time*, we have grown accustomed to this way of thinking that is predicted on looking at things and people through the prism of an 'in-order-to' outlook (be this to accumulate, go faster, be more accurate, or be more efficient), which in turn means that objects and subjects are seen as useful reserves and stockpiles. His critique still has power (and its logic underpins many other extractivist accounts) because he was quick to see the implications of cybernetics as it impacted on the intellectual landscape from the 1940s onwards. Famously in an interview with *Der Spiegel* magazine in 1966, when asked what had taken the place of philosophy, Heidegger responded, 'cybernetics' (Zimmerman 1990). Combining well Heidegger's laments on industrialisation, the science of feedback and use of human interiority as a standing reserve, Vattimo summarises, saying that 'technology, with its global project aimed at linking all entities on the planet into predictable and controllable causal relationships, represents the most advanced development of metaphysics' (1988: 40). Noted before the World Wide Web (1992), the social graphs of Facebook (2005), and in-world datafication of Microsoft and Meta (2021), the lament and remark still has currency.

Backward Note on Heidegger's Pessimism

Note should be made of the physiognomy discussion in Chapter 3, which identified later physiognomy as not simply about physical profiling to infer disposition, but the subsequent *rejection* of essentialism in early Lavater-inspired physiognomy. Spengler's overtly racist and anti-Semitic physiognomy drew on German neoromantic ideas (especially Goethe) of prerational knowledge, lived experience, and intuition, to engage with the vitality of human being that animates the body, rejecting the gaze and logics of systems and geometry. Heidegger's pessimism has similar, if not the same, roots, especially given that Heidegger's well-known anti-Semitism was not only a personal antipathy but one that, through Heidegger's *Black Notebooks*, contaminated his philosophy (Mitchell and Trawny 2017). Yet, his perverse romanticism of social decline is difficult to ignore because of the historic intersection with cybernetics, what he saw as devastation of being, the demise of metaphysics, the establishing of a new order (modernity), distributing and sectioning, making surveyable, turning the world into a causal orderable picture, making humankind machinic, turning humanity into potential and reserve, complete calculation and objectification of the globe, and its conversion into goods and values.

NISHIDA AND THE KYOTO SCHOOL

Having considered energy usage and automated empathy, extractivism, and the critique of Heidegger, we turn to Japan, the Kyoto School, and its novel ways of approaching extractivism. Admittedly not an immediately obvious route for Western readers to consider the environmental impact of automated empathy, the reason for this path is that extractivist and objectification critiques acquire additional power when considered through Japan's philosophical and lived practical orientation to wholes rather than parts. This is especially so given disjuncture between ephemerality and materiality, where experience of digital technologies and content belies complex material dimensions of production. To be clear, despite unique philosophical history and tendencies, Japan's technology policy is familiarly liberal. For example, Japan's Council for Artificial Intelligence Technology Strategy, published by the Japan Cabinet Office, advocates three core values of (1) dignity, (2) diversity and inclusion, and (3) sustainability (Cabinet Office 2019). It is also internationalist, with Japan's initiatives recognising and converging with OECD and UNESCO international policy conversation on AI ethics (Expert Group on Architecture for AI Principles to be Practiced 2021). Indeed, the report of Japan's Expert Group on AI Principles draws extensively on reports from the United Kingdom government, European Union, United States, and international groups such as IEEE (especially its P7000 series of standards, with one of these P7014 that is explicitly about 'Emulated Empathy'; McStay 2021). Notably too, despite what is here argued to be valuable insights from traditional and more modern Japanese ethical thought, regional experts see tendency for data protection to be framed as pro-economy law rather than pro-rights law (Murata 2019). However, despite policy and pro-innovation rhetoric similarity with other regions, there are philosophical differences which provide valuable normative standpoints for approaching development in relation to the full context of human and environmental costs and benefits.

There is homogenising risk in generalisation, but Japan's main belief systems, Shintoism, Confucianism and Buddhism, all speak to a need to begin from the 'whole' as opposed to parts and particularities. Similarly, in addition to the holistic outlook, it is also an 'anthropocosmic' outlook (Tucker 1998) that dethrones humankind from the centre of existence. Eco-systemic by default, it lends well to holistic calls regarding the entire social, technical, and material nature of AI systems (and automated empathy therein). Rare in its ethical pluralism, UNESCO for example observe that AI actors should play an enabling role for 'harmonious and peaceful life', one that recognises 'the interconnectedness of all living creatures with each other and with the natural environment' and belonging 'to a greater whole' (2021a: 8). When one considers for example the life cycle of AI objects, such as an Amazon Echo device, from mining of its materials, smelting and refining, component manufacture, assembly and labour, global distribution and transportation, human labour and training of AI data sets, power to train systems to be more accurate, supportive internet infrastructure (e.g., cables and server farms), shipping of abandoned devices and e-waste, rescue of valuable parts by global poor, and geological impact (Crawford and Joler 2018), there is moral need to be eco-systemic, which is to default to a big picture holistic perspective.

The Kyoto School's starting point of wholes before parts is normatively valuable in this regard, with this chapter arguing that human rights and other global/regional governance organisations would do well to better appreciate ethical traditions outside of liberal thought.

The founding philosophers of the Kyoto School were all connected with Kyoto University. Established by Nishida Kitaro[5] it became a known identity and outlook in the early 1930s (Carter 2013). Although Japan had closely engaged with Western thought since Japan's Meiji Restoration in 1868 and its initiation of a transition to respond to Western dominance, the Kyoto School interpreted and advanced Western ideas to account for Japanese thought. This in turn furthered philosophical bridges between the West and Japan, hybridising several modes of Western thought with longstanding philosophies of Shintoism, Buddhism, and Confucianism. For ethics as they are applied to technology, philosophers such as Nishida rejected suggestion that modernity was a Western experience. Instead, Nishida's self-posed task was to articulate an understanding of modernity that does not originate from the West. With deep interest in Heidegger's idea that there are dimensions of things that are withdrawn from the world yet deeply influence it, in Nishida's (2003 [1958]) parlance this refers to the cosmological self-unfolding 'whole' from which all derives. For Nishida (2003 [1958]) the starting position of wholeness applies regardless of whether one is examining Confucianism (for Nishida including Taoism), Shintoism, or Buddhism. The goal is to grasp life itself rather than particularities. The connection of wholes with environment and climatic logics will be immediate, but it is notable that Nishida and other philosophers associated with the Kyoto School use, and are comfortable with, the language of contexts, backgrounds, and even networks.

While the 'big picture' holistic starting point certainly applies to climate and sustainability concerns (i.e., what is the real and full cost of a given system?), the need to start from the whole has more unique application for AI, automated empathy, and emotion recognition interests. As detailed in Chapter 2, contexts provide meaning. For Nishida, using art as a vehicle for discussion, while Western thought and representation may be analytical and exact in form, it lacks acknowledgment of the formless inexpressibleness from which form is drawn. To link this back to the core interest of the book, automated empathy, there is a key lesson: the extraction of data from experience does violence to it. What was perceived as valuable is extracted and the rest discarded in creation of hyperreal certainty of emotion (divorced in meaning from that which it was derived). Indeed, this point is backed up by empirical work cited in Chapter 2 which used unsupervised learning techniques to track patterns of emotion and correlate behaviour and expressions (Azari et al. 2020). Patterns of the hyperreal sort could not be detected by this inductive whole-first approach, because this included signals and data found in real life. For Nishida (who was very well-versed in Western philosophers of experience, such as Henri Bergson) this distortion is a falsification of experience for the sake of expedience. This ethical orientation is an ongoing respect for the totality (of infinite nothingness) that this takes place within.

What is good, then, is that which acknowledges the context of an action, or what is phased here as a *context imperative*. We can define this as normative instinct that demands responsibility to the bigger picture. What is evil in Nishida is dissociation and forgetfulness of the whole from which we derive. This is because estrangement leads to seeing others as either obstructions to desires, or 'material-at-hand' for individuals to exploit as they choose. Although this is reminiscent of Kantian autonomy and non-interference, Nishida also notes something of the character of lived estrangement, 'as we deem ourselves alone in an alien and mostly inert and uncaring universe' (Carter 2013: 48). A recommendation that flows from the context imperative is that global aversion to dehumanisation through objectification and decontextualisation (as is the case with hyperreal emotion) should recognise Japanese respect for the whole from which particularities are derived.

Practically then, if satisfying other ethical concerns regarding automated empathy, machine decisions and judgements about people, might refer and 'clearly allude' to the limits of contextual understanding by which judgements are made, so the background to the decision is clear. To an extent this is a restatement of universal interest in decisional transparency for governance of AI and big data processing, but it receives extra moral force due to Nishida's and arguably Japan's instinctual aversion to dissociation and forgetfulness of wholes.

COLLECTIVIST VALUE IN ADDRESSING EXTRACTIVIST HARMS

Ethics for Watsuji Tetsurō are found in the 'in-betweenness' of people through communitarian interconnectedness. This is a fundamental point: broadly anti-Western, ethics for Watsuji are primarily about connections and context, not the self and egoism. To lose touch with interconnections is to lose what a person is, because they are divorced from both others and the wider space/time of the shared world. As noted in Chapter 4, this relational view sits very well with African Ubuntu ethics that similarly sees that the individual exists because the group does (Jecker et al. 2002), lending international heft to collectively minded technology ethics.

While respect for context does not equate to conformity, the space/time of the shared world creates the foundation for ethics. This should be seen in terms of *ningen*, that Watsuji (1996 [1937]) states is neither society nor the individual human beings living within it, but the dialectical unity of these two characteristics. Thus, anticipating structuration theory in sociology, while *ningen* is about communal existence 'as the interconnection of acts' it is also about the individual 'that acts through these connections'. Simplified, for Watsuji, Western thought was mistaken in its emphasising of the individual at the expense of all the interconnections in a specific context, time and place that contribute to making a person what they are. Other Japanese philosophers, such as Tanabe Hajime, make similar observations, seeing individual lives as highly fragile, yet when humankind are conceived as a net of individuals and an ensemble of connections, people are strong (Carter 2013: 75).

Temptation to see this as conservative collectivism should be resisted as the point is quite different. As Nishida (2003 [1958]) argues, for society to advance, it must be challenged by individuals whose expressions are society (the collective) challenging and growing itself. Dynamic rather than conservative, the many and the one challenge each other, contradict and grow, arguably in a positive feedback loop, to reach new forms of equilibrium and poiesis. Yet, taking aim at Western phenomenology, individual moral consciousness, Heidegger, and the Cartesian ego, Watsuji (1996 [1937]) problematises the beginning of inquiry from individual consciousness itself. Echoing the context imperative and will-to-holism pointed-out above, Watsuji urges appreciation of in-betweenness, and that which orients and connects people. Deeply ecological, refuting individual consciousness as the basis of ethics, ethics for Watsuji are primarily about connections and context, not the self and egoism. The presence of Confucianism in Watsuji is clear, confirmed by Wong (2012) who notes Confucianism's emphasis of harmony between individuals and their community (Wong 2012). Watsuji's *ningen* is readily and rightly scalable to a principle of community that is much larger than human communities, as he does in his philosophical study of climate (*fūdo*) that de-centres 'man' [sic] from human existence, to better recognise the role of environments.

Climatic Self-Comprehension and Connectivity

For Watsuji (1961) climate is about weather, but it is also social, meaning that it includes the natural environment, but is expanded to also include clothing, agriculture, housing style/architecture, food, recreation, and cooking. Climate then has a discursive character, being a set of inter-linked events, things, and behaviours that form the background and habits for our experience of the world. For Watsuji, 'climate is the entire interconnected network of influences that together create a people's attitudes' (Carter 2013: 132). Watsuji certainly has the natural environment and cultural responses to this in mind, but also human-made infrastructure that connects people (such as transport, postal services, technology, and media). Watsuji would certainly have been fascinated by automated empathy, not least because of the morass of material processes and connections that underpin it, but also because of its potential impact (for better and worse) on attitudes and consciousness, and how these are affected by technologies. Whereas extractivism of the influential Heideggerian sort saw existence solely in terms of people ('man' [sic]), rather than in terms of duality with environments (broadly conceived), Watsuji (and many others in and around the Kyoto School) sees people and the formation of things in much larger contexts.

A famous example is Heidegger's (2011 [1962]) hammer that is used to illustrate how affordances come into being through purposeful action and progression from the background status of object to foreground status of thing. Keeping concern of extractivism in mind, according to Heidegger we should not understand the world as a totality of entities but as set of relationships 'constituted by and for human existence, a structure that enables entities [such as hammers] to manifest themselves or "be" in various ways' (Zimmerman 1990: 140). Watsuji recognises the 'for a purpose' idea, adding that purpose-relation derives from human life and its connection with climatic limitation of human life. He says, 'Shoes may be tools for walking, but the great majority of mankind could walk without them; it is rather cold and heat that make shoes necessary. Clothes are to be worn, yet they are worn above all as a protection against cold. Thus this purpose-relation finds its final origin in climatic self-comprehension.'

Yū (2019) expands on this, discussing Watsuji's lectures on Heidegger. Here, Watsuji queries Heidegger's starting point that the being of technologies come to be through the human concern that manipulates things and puts them to use. Giving examples of clothes or houses, Watsuji's argument is of people responding to what is encountered in the world or climate. Similarly, it is questionable whether a river really is first seen as means to cool a factory, especially as the person who stumbled on that usage may be the rare and subsequently rich exception (most might see it to cool feet on a hot day!). Watsuji also questions 'concern' as a starting point, asking what of enjoyment. A wool-lined coat may keep one warm, but it also feels good. Sensation and a more positive disposition overlaps with Japan's arguably less bleak default position on technology in everyday life. Yū (2019) finally notes Watsuji's interest in mood (*Stimmung*) that for

Heidegger plays a framing function on how things are disclosed for us, but Watsuji takes a climate-based view in that the mood (that discloses what may occurs next) does not simply come to be but is co-created from environment and factors therein. The lesson of seeing a person as embedded in their environments is again the context imperative, one that urges for full and holistic consideration of interests and costs.

CONCLUSION

This chapter began by assessing the increasing material costs of high compute/power systems that derive meaningful information from digital images, such as faces, and natural language technologies that apply to the written word and voice. Here CO_2 increases in relation to both granularity and accuracy of a system, with accuracy being found to be especially problematic for automated empathy, because aspects such as hyperreal emotion sensing is built on weak methodological foundations (absence of consensus about the nature of emotions). Bracketing out other concerns, there is merit in 'inexact' approaches. The next move was to understand material dimensions of automated empathy in relation to extractivism, which is highly Heideggerian due to his apprehension of cybernetics (and its negative implications for knowing in the humanities), but also his in-order-to reading of natural resources (*enframing*) and technologies themselves (*disclosure*).

However, like modernity itself, human rights that inform technology ethics cannot be both universal *and* Western. There has long been tension in questions of universal human rights in that they may not be universal. Quite arguably, they are inextricable from the West's own intellectual and moral history (the Enlightenment) and religious values (Judeo-Christian belief). Moreover, while human rights are often held to be natural, they may also be political values that societies choose to adopt, or values forced upon regions by elites (Dembour 2010). To assess extractivism in automated empathy, the second half of this chapter opted for a pluralistic approach (Ess 2020), that involves scope for the same ethics to be meaningfully adhered to for different cultural reasons (Hongladarom 2016). However, there was value to be found beyond diversifying how the same ethic is reached. Philosophers in and around the Kyoto School advance and upend Heidegger's ideas. Watsuji, for example, offers strong critique of Heidegger's subjectivism, instead focusing on questions based on connections and human ecologies. Attention to 'wholes' that encompass things, people and environment is symptomatic of a Japanese tendency of starting from the indistinct total aspect of a problem, rather than parts and specifics. This is an ecological observation that has application to understanding the full material costs of large-scale automated empathy systems. Yet what was established, especially through Nishida and Watsuji, was a broader environment and holistically motivated 'context imperative' that challenges the intellectual disjuncture between experience of technology and the means of its production. This imperative is a moral force in that dissociation and forgetfulness of the whole can do extractivist violence to that whole. The context imperative argued by this chapter is a ready ethical heuristic, one allowing for consideration of full range of nested contexts, be this material and climate-based, human exploitation, nature of training data for bias in/bias out, impact on marginalised groups, or decisional, in that the contextual elements by which a decision (such as labelling a person angry) was arrived at are always missing. In general, the context imperative is a moral force to address what is lost from the whole in processes of reduction, extraction and production of hyperreal emotion and subjectivity.

Applications and Implications

Positive Education

COVID-19 was a reminder that life-changing crises can come from seemingly nowhere, upending young people's lives along the way. In the context of education, it forced the question of how education might be done differently when in-person interaction was not possible. Online platforms played a vital role in maintaining connections with schools. Universities similarly underwent a dramatic pivot to online platforms such as Microsoft Teams, Google Classroom, and Zoom, a consequence being that Educational Technologies (EdTech) were introduced and normalised without proper scrutiny. Even before this, there was widespread belief that EdTech will fix anything perceived to be broken in education. Speed and interest in EdTech adoption has not been matched by diligence in procurement details, such as who assesses realities of these services against marketing claims, what the criteria for failure are, who has requisite understanding of how data about students and children is processed, and who judges the merits (or not) of using child data to train EdTech AI systems (Digital Futures Commission 2021, Hillman 2022). Concern about questionable procurement practices is not about blaming schools but recognising the burden of having to understand systems whose functioning are opaque. Given generalised lack of technical expertise to understand the specifics of data processing, or how to challenge anything found to be untoward, certainly in terms of the early stages of COVID-19, what could schools have done in such a fast-moving situation? Nevertheless, new modes of education were introduced without scrutiny.

Automating Empathy. Andrew McStay, Oxford University Press. © Oxford University Press 2024.
DOI: 10.1093/oso/9780197615546.003.0006

In addition to merits and data protection concerns of these platforms, the necessity of these platforms also revealed the role that in-person empathy plays in education. Platforms certainly enabled connectivity, but they revealed an empathy deficit too. Teachers and lecturers worked hard to adjust what is usually in-person and 'embodied' teaching to the platform-based environment, but it was clear to most if not all that key factors were missing. Reduced sense of co-presence, cue reading, mutual intelligibility and collaborative feeling-into were reduced by the rapid shift to online teaching. This empathy deficit provided points of opportunity for EdTech enterprises small and large.

Both commercial providers and EdTech champions within schools want to see 'educational technology and digital education prosper as forward-looking optimistic areas of practice' (Selwyn 2021). Indeed, although this chapter has misgivings, EdTech can help with issues of access to education, student academic performance, novel understanding of topics, tailored content, and new approaches to education, all potentially helping address inequalities and improve public education. Here there is a compelling argument to be made for increased use of EdTech in education, especially when considered in terms of access and international inequality. Tailored tutoring apps and systems have scope to vastly improve life chances through education, especially in regions where formal schooling is difficult. This includes for example problems of distance from schools, violence in conflict areas, gender discrimination, costs of materials, overcrowding, poor curricula, and inadequate teachers.

Imperative of access to technology for teaching is amplified through the international decline in qualified primary school teachers, particular in sub-Saharan Africa, and challenges of poverty, gender opportunity disparity, disability, and other dimensions of marginalisation (UNESCO 2021a). Sonia Livingstone, perhaps the foremost scholar of children and the Internet, observes that 'as the Internet becomes ever more embedded into children's lifeworld in a host of increasingly taken-for-granted ways, research is called to examine children's engagement with the world not only *on* but also, more importantly, *through* the Internet' (Livingstone et al. 2018: 1117). This is a hybrid view, meaning that the question is less about whether technologies and data should be involved in education, but asking which technologies, what the data is about, and how systems may be made to work for children's best interests.

Yet, concerns about automated empathy in education have already been shown to be well-founded, including mental privacy, impacts of discriminatory systems, negative impact on wellbeing, lifecycles of data, commensurability of private and public interests, oversight, and whether certain systems do what they claim. Internationally, UNESCO (2021a) warns against use of EdTech, making the hybrid point that 'While there are gains in terms of knowledge and the visibility of educational processes, the growth of machine learning technologies risks fragmenting educational processes into 'data sets' and accelerating trends towards managerialism, surveillance and the de-professionalization of teachers' (UNESCO 2021a: 88). Indeed, on face expression and identification technologies that explicitly fall under the remit automated empathy, arguably with China in mind (examined below) UNESCO states that facial recognition is problematic. It flags that 'the use of facial recognition software and AI for monitoring of students and teachers by states who could use such resources for political surveillance is antithetical to Article 26 of the UN Declaration of Human Rights which affirms the goal of education as advancing fundamental freedoms and human rights' (Ibid.). In the European Union, AI systems in education are already considered high-risk (and that involving emotion recognition very likely to be prohibited), since outcomes may determine educational, professional, and wider life chances. Notably, Recital 35 of the EU's proposed AI regulation leads with concern over 'improperly designed' systems, violating rights to education and training, and the right not to be discriminated against through historical patterns of discrimination (European Commission 2021). Connected, the controversial nature of automated empathy is such that, in 2021 (prior to the release of the draft regulation) the Council of Europe under Convention 108 stated that 'Linking recognition of affect, for instance, to hiring of staff, access to insurance, *education* may pose risks of great concern [sic], both at the individual and societal levels and *should be prohibited*' (2021: 5 [my emphasis]). The 2023 draft of the AI Act signals agreement, stating that use of emotion recognition in education (and law enforcement, border management, and the workplace) should be prohibited (European Commission 2023).

Already a topic of increasing social and international policy attention then, this chapter on automated empathy in education has three interests: (1) the role of emotion and attention tracking through remote learning platforms and in classrooms; (2) how to disentangle what is acceptable profiling in education; and (3) to consider these issues in context of recent updates to the United Nations Convention on the Rights of the Child (UNCRC). Interest in the latter has two principal dimensions, in part because it is international child-focused governance, but also due to its attention to balancing access to education with rights such as privacy.

EMPATHY IN THE CLASSROOM

Given that the focus of this book is on applications of technologies claimed to simulate empathy, and that empathy is intrinsic to being a good teacher, it is useful to have a rounded sense of the role of empathy in education. Empathy takes on multiple meanings in a classroom context, involving interpersonal, prosocial, and interpretive forms. Key to the practice of teaching, interpersonal empathy is a basic perspectival fact of how most people interact with each other. With intellectual origins in David Hume (1896 [1739]), interpersonal empathy has instinctual properties in that it is the wakeful acts of reading, judging, emulating, and reacting to what are perceived as the emotions, intentions, and dispositions of the other. Instinctual character is joined by a prosocial communal character, as argued by the phenomenologist Edmund Husserl (2002 [1952]) who uses the language of 'co-presence' and 'universal Ego-being' to bridge the gulf between subjectivities. In plainer terms, this can be interpreted as 'I have a sense of self and I'm sure you do too, so I know something of what it is like to be you'. On the wider civic goals of education, especially when seen through a Humboldtian prism that champions general learning and cultural knowledge (Peters 2016), empathy has other origins in prosocial political empathy, making it a necessary fact of morality (Kant 1983

[1795]). Indeed, for Adam Smith (2011 [1759]) the moral life of society is underpinned by empathy through its capacity for projective understanding. Notably, to develop qualities of prosocial behaviour in citizenry, in turn, involves a high degree of human empathy (Cooper 2011).

Beyond interpersonal and prosocial dimensions of empathy in education, empathy is important because of its interpretive qualities (Lipps 1979 [1903]). Interpretive and imaginative empathy allows students to 'feel-into' places, pasts, presents, and futures, and even objects. This includes inspirational in-person teaching, but also assistance from technologies to create awareness, represent, and generate affective understanding of situations, things, and processes. Empathy then has aesthetic and sensational components based on projection and feeling-into, which can be split into (1) historic, (2) contemporary, and (3) future parts; with the first being *recreative* in character, the second being about *bridging difference*, and the third being creative to *feel-forward* into how people engaging with designed things will feel. Collectively, empathy enhances and facilitates aesthetic and social experience, the reality of others, and indeed provides a sense of where others are coming from.

Datafied Social and Emotional Learning

Directly and indirectly, empathy informs social and emotional learning, which is 'the ability to understand, manage, and express the social and emotional aspects of one's life in ways that enable the successful management of life tasks such as learning, forming relationships, solving everyday problems, and adapting to the complex demands of growth and development' (Elias et al. 1997: 2). Complementing academic skills, school-based social, and emotional learning aims to develop young people's ability to relate and identify, communicate, make decisions, and nurture self-esteem. Although social and emotional learning in its modern form is around 25 years old, scholarship in social and emotional learning can be traced back to 1900 through social inequality policy problems (Osher et al. 2016). In its original guise social and emotional learning was not overtly

politically partisan, although there is a clearly a liberal flavour to its principles of self-direction, self-control (rather than external control), and
inculcation of social competence. Reviewing 100 years of social and emotional learning, Osher et al. also recognises 'affective education' that, from
the 1970s, sought to address students' feelings about themselves, affective
states, and emotional competencies. In addition to the student individual
and their introspective experiences, social and emotional learning-related
scholarship later became interested in interpersonal, relational, and dynamic understandings of emotion in social groups, but also how student
individuals will attune and negotiate these group contexts.

Social and emotional learning is then the attempt to balance cognitive elements in education (knowledge acquisition, analysis, reasoning,
and memory), with management of feelings and emotions, perseverance to achieve goals, and to better work with others. This is something
that influential bodies, such as UNESCO, see as 'fundamental to human
creativity, morality, judgment, and action to address future challenges'
(UNESCO 2021, 68), with the Organisation for Economic Co-operation
and Development (OECD) also seeing utility in *measuring* child sociality
and emotion, to help create work-based skills and disposition for emergent economic realities, especially those involving autonomous systems
(OECD 2015). This introduces new factors beyond a holistic view of child
education, involving strategic shaping of children for labour and the mar-
ketplace. To reinforce the point, the OECD (2018) elsewhere observes that
social and emotional skills, especially those connected with grit, resilience,
perseverance, creativity, growth, emotion regulation, self-management,
and control, are seen as critical to the workforce of the future.

The turn to measurement of emotion is crystallised by the World
Economic Forum, that champions use of technology to assist in advancing
social and emotional learning in schools. Citing the facial coding of
Affectiva to 'recognize, interpret and simulate human emotions' the Forum
uncritically celebrates it as showing 'great promise for developing social
and emotional skills such as greater empathy, improved self-awareness
and stronger relationships' (WEF 2016: 15). Alongside these behavioural
and economic interests is that the construction of systems and tools to

measure qualitative states and child interactions serves policymaking, rather than children themselves. This in part is about the belief that data about sociality and emotion may providing the evidence base through which decisions and successes of social and emotional learning can be judged (Osher et al. 2016), but it is also part of a larger meta-narrative of building big systems in education. Critical EdTech scholarship is alert to this, with critical EdTech thought-leader Ben Williamson observing that datafication of emotion serves the overall assemblage of technology-focused education policymaking rather than children (Williamson 2021a). This assemblage involves questionable levels of datafication, but more broadly a policy infrastructure that for Williamson is constituted by organisations, expert knowledge, concepts, techniques, and technologies, all built on datafication that does not sufficiently question negative implications of this datafication. In addition to serving of bureaucracy and policy, Williamson's next concern is that quantification of qualitative dimensions of child life ultimately serves industry rather than children, especially given the narrow market-based interest in child work-based skills and disposition for the workplace.

As discussed in Chapter 2, bridging of quality (emotion) and quantity (numeric data) has roots in Manfred Clynes' 'Sentics' that sought to provide a cybernetic understanding of emotion. This would help children 'be in touch with their emotions' and, echoing prosocial empathy principles above, allow 'different races and backgrounds to experience their common basis in humanity' by being sensitive to the emotions of others (1977: xxii). In 2001 this theory began to take tangible form as Rosalind Picard (the originator of the term and practice of affective computing) recognised the scope for emotion-sensing in education. Indeed, her team traces the development of computerised tutors that are sensitive to emotion and affect back to 1988 (Lepper and Chabay 1988; Picard, Kort et al. 2001). Noting the diversity of emotional and affective behaviours in the classroom (frustration, dejection, hopelessness, curiosity, interest, exploration, and enjoyment), Picard et al. (2001) sought to build a 'computerized learning companion that facilitates the child's own efforts at learning'. The goal of the companion was to improve pedagogical techniques by using computer vision techniques to watch and respond to the affective states of children. The context of this observation was twofold: that the interplay between emotions and learning is highly important; and that education systems do not focus sufficiently on *how* rather than *what* students learn (Picard, Kort et al. 2001).

More recently, Sidney D'Mello's 'Affective AutoTutor' detects and responds to learners' boredom, confusion, and frustration. It does this through facial coding, tracking of interaction patterns and body movement (D'Mello and Graesser 2012), providing motivational feedback to students through appropriate facial expressions and voice emotion. D'Mello (2017) also details voice-based work (Forbes-Riley and Litman 2011) that in addition to identifying correctness and incorrectness of given answers also detects learners' certainty or uncertainty. Other work focuses on attention, rather than emotion, with D'Mello (2017) accounting for tracking of students in a classroom using cameras placed around the blackboard area of the classroom. Computer vision techniques gauged head detection and head-pose estimation, which were then used to detect student attention, which was subsequently validated via self-reports from students. Mention should also be made of teachers, as their teaching methods may also be subject to analysis. This entails recording of in-classroom audio content and use of automated means to predict the level of discussion in these classes.

Although conceived with student well-being, and teacher reflection and improvement in mind, risks are chilling effects, tone policing, punitive performance management, and personality management of children based on *hyperreal* and *unnatural mediation*. Biopolitical, they also raise clear *mental integrity* questions regarding conditioning and impact on child subjectivity, more so given the presence of both surveillant and hyperreal emotion. This is an articulation that filters emotional life through hyperreal filters of economic, psychological, statistical, and calculative expert frames of what emotional life is, using APIs and attractive data visualisation to hide the truth that what emotions are is unclear. This intellectual vacuum allows for articulation of emotion to suit any given political agenda, be this neoliberal or conformist. Political agendas of diverse types are served by a biopolitics of emotion ambiguity, one where child neuroplasticity and subjectivity risks being created and funneled through automated empathy. This is some distance away from the aforenoted Humboldtian model of education that champions comprehensive learning and well-rounded citizens who are not defined in relation to work. On the modern neuroplasticity dimension, Williamson (2018) draws on Whitehead et al. (2018) to account for moulding of neuroplasticity for the market as 'neuroliberalism'. This grounds numerous issues, including commensurability of private interests with the public good, commodification of child behaviour, engineering of the brain and responses to experience, shaping of behaviour, and biopolitical interest in augmenting child behaviour in education.

In addition to concerns about child and student shaping are those of an experiential sort. Andrejevic and Selwyn (2020) identify a subtle but significant point regarding facial recognition technologies: that children and students are not 'seen' but processed. In addition to experience of objectification is the experience of having to game automated empathy systems, to empathise with *their* affordances, and how to perform to elicit good behaviour scores. Related to the perverse gamification of subjectivity and performance is the sense of being 'always on' and never being able to mentally drift for concern about what outward features may be communicating. In addition to these innate problems with facial recognition and emotion scoring of children, this practice must be seen in context of their embedding in other systems. Andrejevic and Selwyn (2020) for example point to facial recognition and how (outside of education) it has been used with risk scoring systems. The same applies to education and that what they see as a 'cascading process of automation' would inevitably apply beyond facial recognition to include affect and emotion profiling. Consequently, any suggestion that automated empathy systems would only be used to help teachers improve their teaching should be treated with caution, given a broader history of mission creep with surveillance technologies and educator interest in making systems co-operate in name of efficiency.

APPLIED EDTECH

Recollecting that post-phenomenology recognises that human experience is inextricable from technology (Ihde 2010), postphenomenological interest in education will see technologies for writing (such as pencils) and idea dissemination (such as books) as featuring in the most stripped-back of classrooms. More sophisticated technologies also have a long history. These includes Thomas Edison's vision of film as an educational technology (1920s); Sydney Pressey's 'machine for testing intelligence' (1920s); and Burrhus Skinner's mechanical and electronic 'teaching machines' that, following logics of operant conditioning, sought to speed-up student feedback (Watters 2014, Selwyn 2017). Education then has always had a mediated and hybrid character. Today, start-ups see opportunity, and established technology firms are expanding into education. The UK's Centre for Data Ethics and Innovation (2021) summarises these EdTech interests as

- *Predictive analytics*: gauging performance, skills, likelihood to drop out, potentially through eye-tracking, emotion, and other affective states.
- *Aggregated data on students' attainment and behaviour*: that assists in short-term planning for teachers and long-term strategy for schools.
- *Natural language processing*: that may assist in marking written work, languages teaching, and monitoring of digital classrooms.
- *Automating routine tasks to lessen teachers' workloads*: including compiling and distributing learning materials, marking homework, and monitoring student attainment.
- *Personalised learning systems*: that generate recommendations for learning materials and pathways through predictive analytics and real-time feedback.

The commercial appeal of modern EdTech is apparent in that there is monetary profit in 'disrupting' public education, helped by belief among technology-focused education policymakers that the future of education is best served by new and more detailed forms of learner analytics. For example, financing for EdTech start-ups in the COVID-19 pandemic year of 2020 increased from $4.81 billion to $12.58 billion worldwide (Singer 2021). Even before the COVID-19 pandemic EdTech start-ups were growing quickly, but as schools switched to online methods EdTech firms temporarily made their premium services free to teachers for the rest of the school year (Ibid.). The significance of this can be read two ways: one is prosocial; the other is that it is a freemium growth technique, where crisis-hit schools would become reliant and locked-into systems.

Popular technologies include platforms and video services, messaging and collaboration tools, and VR and AR, delivered by adtech behemoths such as Google. Google's 'Workspace for Education' provides a suite of cloud-based tools for schools between kindergarten and 12th grade (K–12), higher-education institutions, and US home schools. Indeed, in 2021, at Google's annual I/O Developers Conference, Google's CEO Sundar Pichai hailed how Google during the COVID-19 pandemic had helped 'students and teachers continue learning from anywhere', indicating Google's future in this area by announcing a new AI language platform to allow students to ask natural language questions and receive sensible, factual, and inter-esting conversational responses. This, however, must be seen in context of Google's business model, which is tracking human attention and adver-tising (Williamson 2021b). One could initially see this as coincidence of interests, but there is evidence of strategic overlap. The US State of New Mexico took out legal action against Google, stating that Google Education had mined web, location, contact, voice, and other behavioural data, also observing the range of sensitive searches that teenagers will conduct on mental health and issues they are too embarrassed about to ask in-person. Although the judge ruled that Google did not violate the US Children's Online Privacy Protection Act, this was a technical rather than moral win for Google. The judge did not find that mining does not take place but ruled that there was no legal requirement for privacy notices to be written in terms understandable by a child under the age of 13 (Needleman 2020). This speaks to a larger point about the interests of technology companies in that data inferences about children and students (including emotions) are used to train the neural networks owned by the EdTech providers for purposes outside of education. Be this legacy companies such as Microsoft or more recent entrants such as Affectiva, all are keen to obtain training data to improve their algorithmic services in other business contexts. The result is that data about child emotion is commoditised to improve algorithmic services, create competitive difference (in terms of how many faces are analysed) and serve business and strategic contexts for which the student data was not intended (such as car in-cabin analysis, retail, out-of-home advertising, market research testing of advertising effectiveness, media, entertainment, and more).

As COVID-19 accelerated use of automated empathy, start-ups were quick to develop apps for online teaching so teachers could tailor instructional activities based on student affective responses, to try and remedy problems of presence, engagement, and teachers' difficulty in 'reading the room' online. For example, under the auspice of *4 Littletrees*, the Hong Kong-based firm Find Solution AI uses face-based emotion recognition to measure 'facial expressions and emotions and generate detailed reports for teachers'.[1] As children study, the system measures muscle points on student faces via the camera on their computer or tablet, classifying expressions into basic and hyperreal emotions, including happiness, sadness, anger, surprise, and fear. The system then generates reports on students for teachers and parents, forecasts exam performance, and makes recommendations on remedies for struggling students. Similar, *Dreambox*[2] is a US-based K8 adaptive learning app that tracks students and adjusts online Mathematics lessons if students are struggling, not engaging, getting frustrated, or are inferred to be bored. This involves gauging of emotion and profiling in reference to wider information about a student. *Dreambox* also judges performance, provides feedback, and through prompts modulates the behaviour of students to goals given by a system. Other start-ups such as Proctortrack promise to automate overseeing of exams processes, where 'Each student will provide face, ID, and knuckle scans, which will be measured against the student's baseline biometric profile, stored on file' (Proctortrack 2022).

With Proctortrack integrating with Moodle, Blackboard, Canvas, and other platforms teachers and lecturers will be aware of, the system uses desktop cameras to monitor face, eye, and biometrics behaviour to infer whether a student is behaving suspiciously or cheating. Students who move their eyes in ways deemed suspicious by the system, display the wrong facial expression, or move their bodies wrongly, will trigger a flag for the system (Digital Defend Me 2022).

Larger platforms also see opportunity in automated empathy, with the education branch of the global technology company Intel stating that they are researching how recognition of affect may personalise learning experiences and provide personalised learning. Key for Intel is to collect 'multiple points of input and output, providing analytics in real-time that lets teachers understand student engagement' (Intel Education 2022). The idea is multi-modal, involving three inputs to a classroom computer that both records and predicts engagement during a class session. These include *appearance* where cameras extract facial landmarks, upper body, and head movement and pose; *interaction* and how the student uses input devices such as keyboard and mouse; and *time to action*, or how long is the student taking to complete tasks or act on a learning platform. Intel have also collaborated with Classroom Technologies to develop software that works with the platform Zoom.[3] This is to identify information about engagement and participation, such as how often and long a student speaks in class. Analysis of facial expressions in relation to what students are working on is intended to provide contextual understanding about student performance, attention, and who needs help or greater challenge. Microsoft has also long offered facial coding products to schools, marketed alongside their face identification products under the auspice of Azure, their cloud-based services (Microsoft 2022). In 2022 Microsoft promised it would discontinue emotion recognition through Azure, which was widely interpreted to mean that Microsoft would desist from all work on emotion recognition. However, this is not what they said. Discontinuation applies only to the Azure Face services, with Microsoft adding that they 'need to carefully analyze all AI systems that purport to infer people's emotional states' (Crampton, 2022). This is a much weaker statement of intent than one that says Microsoft has stopped all emotion recognition development.

Indeed, scope to mediate emotion is important to Microsoft's mixed reality and metaverse ambitions. Some care is required here in that nonbiometric mixed reality (through VR and AR-based systems) *will* certainly create prosocial pedagogical opportunity for novel means of learning and understanding (Daniela 2020). Increased usage of immersive learning has scope to enrich and make tangible the impact of climate change in faraway places, of biological systems, or of historical situations (Microsoft Education 2022). I argued similar in *Emotional AI* (McStay 2018), suggesting scope for 'affective witnessing'. This means that VR and mixed reality grants not only cognitive and intellectual comprehension of people, places, periods, fictional creations, objects, and systems, but also facilitates aesthetic, kinaesthetic, and affective understanding.

However, commercial interest in the metaverse is not only about immersive understanding, interaction and overcoming of spatial distance, it is also biometric. Staying with Microsoft, their 'Mesh for Teams' portal renders bodily behaviour and facial expressions for in-world avatars. While aimed at the world of work (Roach 2021) the overlap with education is clear given the now fundamental role of the video platform Teams in education. Given both Microsoft's history of workplace analytics to help managers understand person metrics, worker busyness, productivity, effectiveness and habits (Microsoft 2021), and the datafication of education as observed by Williamson (2021a), the risk is unwanted surveillance of bodies (both real and digital representations) in the name of social and emotional learning. This book reasons that modalities of mediated empathy should be restricted to prosocial and interpretive sorts, rather than that based on directly judging and profiling student emotions.

PSEUDOEMPATHY

UNESCO's 2021 report on the new social contract for education is notable, cautioning against fetishisation of 'digital knowledge' in the post-Renaissance West. It remarks, 'This side-lining of nontechnology ways of knowing has deprived humanity of a vast and diverse archive of knowledge about being human, about nature, about environment and about cosmology' (2021: 36). Their call for nondigital ways of knowing is not just about what is studied, but pedagogy as an 'expert competence' that rejects or is suspicious of 'informal, indigenous, and not easily accessible knowledges' (Ibid.). If automated empathy in education was to meet scientific standards of accuracy and it was to have verifiable prosocial benefit in education though increase in wellbeing and performance, there would be scope for debate here. However, despite a veneer of positivism, it is digital and automated profiling that is innately questionable, especially due to the inherent ambiguity of emotional life. As detailed in Chapter 2, the automated empathy industry is *still* dependent on Ekman and Friesen's (1971) hyperreal basic emotions model. While leading industry figures recognise its limitations, such as Microsoft Research's Daniel McDuff and Mary Czerwinski (2018), this has not stopped Microsoft from using this approach. This is evident in their Microsoft Azure's Face API pages selling their face and facial feature recognition services, with emotion labels being happiness, sadness, neutral, anger, contempt, disgust, surprise, and fear (Microsoft 2022).

Problems of ambiguity, method, and inaccuracy are amplified by observation that facial coding is especially poor with children, due to their immaturity and lack of development in emoting (McStay 2019). Barrett et al. (2019) state that, 'In young children, instances of the same emotion category appear to be expressed with a variety of different muscle movements, and the same muscle movements occur during instances of various emotion categories, and even during nonemotional instances' (2019: 27). Giving the example of fear they further point out that young children's facial movements lack strong reliability and specificity as an indicator of emotional experience. Whereas in facial coding systems fear equates to a wide-eyed and gasping facial expression, this correlation of expression and experience has rarely been reported in young infants. Although one may conceivably try to adjust the system to admit of an averaged set of expressions that children display in relation to named emotions it is notable that even large providers such as Microsoft Azure pages make no reference to the unique emoting behaviour of children, leaving one to assume that adult oriented technologies are being used with children.

Yet, despite the rightful critical din of claims of pseudoscience, this is not the core problem. Academic researchers of mediated emotion, unencumbered by need to develop standardised industrial products, recognise the inherent complexity. This takes the form of understanding that people have an enormous socially contingent repertoire of emotions, the role that context plays, the value of multimodal sensing to build confidence in results, situation appraisal, expressive tendencies, temperament, signal intent, subjective feeling states, and polysemy and multiple meanings of expressions (Cowen et al. 2019). The risk of a pseudoscience-based critique is that it invites *more* profiling and granular labelling of brain, bodily and situational interactions. Indeed, writing directly on emotion, technology and education, D'Mello (2017) argues that because affective states 'emerge from environment–person interactions (context) and influence action by modulating cognition', it 'should be possible 'to "infer" affect by analyzing the unfolding context and learner actions'. One can debate whether computational inferences of emotion expression and behaviour is always bound to fail, even with inclusion of multimodal measures, but critical mindfulness is warranted for critique based on inaccuracy and problematic science.

Learning from China

Despite ongoing datafication of education, education policymaker appetite for 'learner analytics', and that numerous start-ups and legacy technology companies are building automated empathy systems for education, actual usage in classrooms and online platforms is currently not the norm. It is useful to consider lessons from China that has piloted these systems and intensively promoted EdTech. China invested heavily in AI for education, especially since the Chinese government published its 'New Generation

Artificial Intelligence Plan' in 2017. Assisted by access to large data sets, funding from venture capital, tax breaks to improve learning, scope to experiment with new technologies, impetus from the COVID-19 pandemic and a competitive academic culture (Liu 2020), this meant that China's EdTech market grew fast, backed by Chinese legacy companies such as the Alibaba Group, international funders such as the SoftBank Vision Fund, and US investment firms (Shu 2020). Chinese firms using emotion recognition technologies for education include EF Children's English, Hanwang Education, Haifeng Education, Hikvision, Lenovo, Meezao, New Oriental, Taigusys Computing, Tomorrow Advancing Life (TAL), VIPKID, and VTRON Group (CDEI 2021a).

However, in July 2021 China's Central Committee and State Council made significant reforms, banning companies that teach school curriculum subjects from making profits, raising capital, or listing on stock exchanges worldwide. The concern was high levels of profit being made from China's competitive education system, (such as online tutors and apps that tailor after-school learning for primary and middle school children). Regional Chinese EdTech analysts point out though that as student homeworking wanes, classroom technologies may even be entering a golden era due to desire to increase in-class effectiveness. This leads Chen (2021) to observe that while China's EdTech reforms were severe, solutions that help analyse in-class academic performance will be in demand.

Hanwang Education, for example, provides in-class analytics to offer weekly scores of students, while Haifeng Education goes as far as tracking student eyeball movements. Notably there is no mention in promotional literature of Chinese regional variations of social and emotional learning (and Western neoliberal interest in workplace readiness). Goals are different, involving the self, but also others and the group. A student is expected to exhibit different forms of self-regulation and behaviour when dealing with others, including collective awareness, and managing of relationships between oneself and the collective (Osher et al. 2016). China's experience with emotion profiling and automated empathy in the classroom is instructive for the rest of the world, especially given that students will feign interest and perform to receive rewarding metrics. The performative and self-policing aspects are key and will also feature later in consideration of automated empathy in the workplace.

A report by human rights organisation Article 19 observes that Chinese pre-pandemic uses of expression recognition in schools tended to fall into three general purposes: conducting face-based attendance checks, detecting student attention or interest in a lecture, and flagging safety threats (2021: 25). These practices are neither popular with students nor teachers due to privacy questions and of actionable feedback from the systems themselves (Article 19 2021). Does inattentiveness to part of a lesson, for example, signal boring content or delivery? Further, the systems offer no suggestion on how to improve these. Thus, while much methods-based critique of emotional AI focuses on the problem of reverse inference (the fallacy that an expression necessarily reflects an emotion), there is the more general question of whether emotional AI in EdTech improves anything. Notably though the report found that these emotion-based technologies resonated strongly with parents, motivated by perceived higher quality of learning. While one might not expect parents to subject their children to surveillance practices and their chilling effects, the competitive nature of China's education system is such that parents do not reject it. This does not mean they are in favour, but potentially ambivalent. Related, our own findings at the Emotional AI Lab found high levels of parental ambivalence in parents and their attitudes to child wellbeing wearables that track affect and emotion (McStay and Rosner 2021). Although education and personal wearables are different, parents in both instances will want to do what they see as the right thing, meaning they are pulled in multiple directions, such as being willing to exchange privacy for perceived safety or educational improvement. One might hypothesise similar about China in that while parents will care about privacy, experience of school, and mental integrity, in an educationally competitive country that is deeply wedded to technology, parents see little option given need to consider the long-term best interests of their children.

AUTOMATED EMPATHY, EDTECH, AND THE RIGHTS
OF THE CHILD

It is useful to consider questions of access to education, privacy, and pro-social and interpretive empathy through the prism of the *United Nations Convention on the Rights of the Child* (UNCRC). Ratified by most of the international community for persons under the age of 18, it is an agreed set of rights and protections. Force of the Convention depends on whether the UNCRC has been incorporated into domestic law. The UK, for example, is a signatory to the UNCRC, but it has not incorporated it. In practice, this means that courts in the UK may be persuaded by the moral force of the UNCRC when making legal judgments, but they are not bound by it.

The Convention details the most fundamental rights of a child (such as parental access, sanitation, nutrition, care, and physical and mental wellbeing), but also states that signatories should act in the best of interests of children, that children have the right to have their privacy protected, and that their lives should not be subject to excessive interference. The Convention's need for adults to act in the child's best interests (Art. 3) is a flexible phrase that risks justification of all sorts of impositions on a child, as suggested in the discussion of parents of Chinese children above. Commentary on the Convention provides useful interpretive detail about 'best interests,' stating that justification for a decision that will affect children (such as the use of automated empathy in education) should be explicit in the factors that have been considered. This is reinforced in §15e of General Comment No. 14 (UNCRC 2013) regarding cases of data collection, which requires 'that the child's best interests are explicitly spelled out'. Notably the Convention does not equivocate on cultural sensitivity. For example, one might argue that norms and values in educational approaches around the world differ in China. Although the UNCRC allows for cultural sensitivity in relation to child identity and issues such as foster home placement, cultural identity may not be used to perpetuate practices that deny the rights guaranteed by the Convention. Further, on education itself, the UNCRC states that 'education is not only an investment in the future, but also an opportunity for joyful activities, respect, participation and fulfilment of ambitions' (General Comment No. 14 2013: §79). In addition to perceived increase in quality of education from Chinese parents, the UNCRC's more expansive view of the forma- tive role of education clearly expands its role from being primarily that of preparing future workers.

Of the 54 articles on the rights of the child there are arguably six *overtly* and *directly* relevant (rather than broader but applicable) articles to digital data-rich education, including: acting in the child's best interests (Art. 3); freedom of thought (Art. 14); privacy (Art. 16); access to education (Art. 28); education that allows children to reach their full potential (Art. 29); protection against all other forms of exploitation (Art. 36), with this latter article included here because digital data from education is used to advance economic interests beyond education. In the case of EdTech companies, even the most well-meaning of business interests are foremostly financial. This puts them at odds with Art. 3 that requires that the interest of chil- dren be the primary consideration. However, practicality is required be- cause there are all sorts of commercial services used in education. Rather, the presence of private interests in delivering public goods should be seen both in context of the technology itself (that may use biometric data about children) and how EdTech that uses emotional AI fares with other aspects of the Convention.

The right to freedom of thought (Art. 14) typically means that a child should be able to have their own ideas, thoughts, opinions, and beliefs. The overlap with *mental integrity* (a concern throughout this book) is clear, which in context of automated empathy refers to use of biometric data to condition people in some way. This is intensified by the right to privacy (Art. 16), which includes a child's 'correspondence' that in a modern online context may mean interception of online communication and personal data. Yet, the right to education (Art. 28) and to reach full potential through education (Art. 29) raises difficult questions, not least that that applications of automated empathy in classrooms (online and in physical spaces) may help rather than hinder a child's ability to reach their 'fullest potential'. The sort of care urged by the hybrid approach advanced in Chapter 4 is helpful, especially as it assesses the terms of engagement between people and technologies. For example, there is an argument to be made in that a child obliged by circumstances to use online platforms may agree that use of cameras to mediate facial expressions to represent emotion expressions on avatars in in-world contexts helps them reach their fullest potential. In a distance learning context where students and teachers may meet in online classes face-based profiling via cameras for online representation (such as through an avatar) is seen here to be acceptable assuming it is *only* for display purposes. The key difference is movement of a person's real face landmarks would not be labelled (with an emotion type) or recorded, meaning that in-world expression mediation would be temporal, of the moment, and not recorded for further analysis (just like everyday interactions). Concerns start when judgements are made about students, especially about their in-class or in-world emotion behaviour, which is argued here to not only fall foul of the Convention (esp. Art. 3, 14, 16, 36), but also rights to education because efficacy is a long way from being agreed.

Argument for automated empathy and labelling of emotion is further nullified if proponents cannot (a) provide sound science about the nature of emotion, (b) suggest meaningful and numeric thresholds to make judgements about a person, and finally (c) offer upfront verifiable evidence that emotion recognition for social and emotional learning works. Beyond effectiveness criteria, there are the broader but applicable UNCRC-based problems, not least its interest in 'mental, spiritual, moral and social development' (Art. 27). This encompasses the experience of being processed rather than seen at school, how young people experience their own emotional and affective states, perverse outcomes of having to gamify automated empathy surveillance, and chilling effects. Broadly, this is the problem of emotional and behavioural self-objectification, self-surveillance, and the internalisation of a self-observer perspective upon child bodies and behaviour. Especially when seen in context of growth in child mental health disorders, the possibility of self-observer perspectives and behavioural chilling effects has even greater reason to be avoided. Notably, in the UK, but with likely relevance to elsewhere, children aged 5 to 16 years with a probable mental disorder were in 2021 more than twice as likely to live in a household that had fallen behind with payments (16.3%) than children unlikely to have a mental disorder (6.4%). Further, given that poorer children are more likely to attend public schools with larger class sizes (NASUWT 2021), and that the impetus of automated empathy in the classroom is to assist overstretched human teachers, it is likely that poorer children with mental health problems are most likely to come under the gaze of systems that profile behaviour, attention, and emotion. Again, argument for emotion-based automated empathy in education is deeply flawed.

Hybrid Rights of the Child

Coming into force in 1990, the United Nations Convention on the Rights of the Child discussed above was designed to apply to all rights of the child. Over the years however it became clear that there was need to explicitly address the digital environment. In 2021 this was remedied by an update in the form of General Comment 25 (UNCRC 2021). Bland in name only it details how child rights in the digital environment should be interpreted and implemented by States around the world. From a hybrid ethics perspective, the tone of the Comment is notable. It seeks to protect children and young people, yet also sees access to digital services as an essential part of twenty-first century living.

As this General Comment was being formulated, the Emotional AI Lab submitted evidence[4], uniquely focusing on data protection harms associated with datafied emotion in education and toys. It appears that we were heard as General Comment 25 contains multiple mentions of emotion analytics (§42, 62, 68), finding this to interfere with children's right to privacy, and freedom of thought and belief, also flagging 'that automated systems or information filtering systems are not used to affect or influence children's behaviour or emotions or to limit their opportunities or development' (UNCRC, 2021 §62). General Comment 25 is not against EdTech, seeing increasing reliance on digital services and opportunities in digital technology for children around the world. While privacy, online abuse, profiling, sharenting (parents posting content about their children online), identity theft and automated data processing (in certain contexts) are recognised as social ills, so is the potential for exclusion from digital and hybrid life, that is, one based on positive relationships with emergent technologies that advance rather than harm mental integrity.

In relation to education itself, the digital environment is seen by General Comment 25 as providing scope for 'high-quality inclusive education, including reliable resources for formal, non-formal, informal, peer-to-peer and self-directed learning' (§99), also with potential 'to strengthen engagement between the teacher and student and between learners' (ibid.). Part XI(A) that deals explicitly with education attempts to balance rights to access and avoidance of greater global inequality, with harm mitigation and avoidance. In trying to find this balance, it urges States to 'develop evidence-based policies, standards and guidelines for schools and other relevant bodies responsible for procuring and using educational technologies and materials to enhance the provision of valuable educational benefits' (§103). It also specifies that standards for digital educational technologies should ensure that child personal data is not misused, commercially exploited, or otherwise infringe their rights, such as documenting a child's activity and then sharing it with parents or caregivers, without the child's knowledge or consent. The point about commercial exploitation is especially significant for this chapter, as commercial exploitation often simultaneously means (1) unethical practices and (2) activities used to benefit commercially from property.

CONCLUSION

Should there be automated empathy in education? As a prima facie principle, the suggestion that technologies may assist teachers in their activity is not at all problematic. The hybrid orientation of this book expects increased use of online technologies, often seeing this as a good thing. These should help increase access, quality and student flourishing (through learning, understanding, thought, creativity and positive relationships with others). Indeed, with protections in place, this chapter does not assume that private interests and serving the public good are incommensurable. Also, modes of mediated empathy based on 'feeling-into' places, pasts, datasets, chemical structures, planets, parts of the body, and other things that benefit from enhanced and affective interpretation are welcomed.

However, on automated empathy in education, most notably hyperreal face-based analytics, to proceed from such weak scientific foundations arguably makes the premise terminally problematic, especially for evidence-based education policy. Expanded, there are numerous methodological and normative problems, each amplified by mental integrity concerns. These include (1) serious questions about effectiveness, validity and representativeness of training data; (2) that financial incentives and the well-being of school children do not align in relation to automated empathy; (3) it is morally problematic to use inferences about children's emotions to train neural networks deployed for other commercial purposes; (4) mission creep, where in-class data may be used for other socially determining purposes (such as social scoring); (5) risk of self-surveillance and chilling effects in the classroom; and (6) data minimisation questions that ask whether automated empathy is necessary for successful education.

This chapter concludes that, in the case of automated empathy and facial coding in the classroom, these technologies are incommensurable with current and near future social values. The only instance here considered acceptable is face-based mediation via cameras for online representation (such as through an avatar), assuming it is for display purposes *only*. The difference is that movement of face landmarks would not be recorded nor labelled with an emotion (so being akin to everyday nonrecorded interaction). Indeed, automated judgements of student emotions in virtual spaces (and physical classrooms) are argued here to fall foul of the United Nations Convention on the Rights of the Child (esp. Art. 3, 14, 16, 36). Ultimately, even with improvements in methodology and efficacy, inescapable and recorded observation of emotional behaviours does not align with need for mental and emotional reserve to ensure human flourishing.

Automating Vulnerability

Sensing Interiors

A 2020 advertising campaign for Volvo by agency Forsman & Bodenfors[1] tells the viewer that when the three-point seatbelt was introduced by Volvo in 1959 it was widely seen as a 'terrible idea'. In black and white featuring testimony from crash survivors, the ad quotes *The New York Times* calling mandatory seatbelt wearing a 'violation of human rights'. Other news reports citing scientists are also quoted, stating that seatbelts would be ineffective. The ad closes with: 'As the next step, we will introduce in-car cameras to prevent intoxicated and distracted driving.' The not-so-subtle suggestion of this strategic communication is that critical press and those interested in the social impact of new technologies are over-reacting, that society will familiarise with new practices, and it will be better for them.

It is not only the carmaker Volvo that is developing in-cabin sensing. Informed in part by practicalities of safety legislation, standards and in-cabin personalisation, Ford, Porsche, Audi, Hyundai, Toyota, Jaguar, and Aisin, among others, are each developing affect and emotion tracking systems. These are to assist with occupant safety and profiling the emotional behaviour of drivers to personalise in-cabin experience. Mercedes-Benz (a brand of Daimler AG) for example speaks of improving the comfort by having a 'butler' attend to driver's needs,

Automating Empathy. Andrew McStay, Oxford University Press. © Oxford University Press 2024.
DOI: 10.1093/oso/9780197615546.003.0007

moods, and comfort, through sensing and learning of preferences (Mercedes-Benz 2021). These legacy car makers are being joined by numerous start-ups, each promising to create value from data about in-cabin occupants.

Divided into two sections, the first part of this chapter examines the significance of cars and an automobile sector that is increasingly feeling-into the affective and emotional states of drivers and passengers. Mostly involving manual driving, this takes the form of cars that contain cameras and sensors that point inwards, as well as outwards, to detect emotion, attention, and behaviour. Given that cars like homes are highly intimate spaces, the proposal to passively surveil occupants is significant. Cars are especially interesting in this regard because in many countries (not least the USA, but others too) cars are icons of freedom and self-determination, linking well with issues of agency and privacy. They are also a good example of longstanding hybridity, entanglement with technology and, when seen holistically, the shaping of human life and urban habitats around the social realities that cars afford. Against this context, the first part of this chapter details industrial development, support from regulation, and outcomes of interest, especially regarding potential for international standards of affect and emotion, and how these connected cars will interact with third-party organisations, such as insurers. The second shorter part of this chapter considers autonomous driving and ambition from Microsoft researchers to model behaviour of the human nervous systems and factor this learning into cars themselves. This in effect is to build safety systems through what this chapter considers as *auto-vulnerability*. A more speculative proposition than the idea of in-cabin cameras and sensors, the idea of equipping cars with something akin a human autonomic nervous system and affective reactions has intuitive sense, making it worthy of inclusion and assessment. Deeply biosemiotic in character, the goal of this is highly autonomous transport that can appraise situations to better plan, react, cope with changing environmental conditions, and reduce risk to the car and its occupants.

METERING SUBJECTIVITY IN AUTOMOBILITY

In addition to practicalities of carrying people and things from A to B, cars act as places of safety, sites of warmth, seats of familiar and repetitive actions, retreats for privacy, means of thrill, acoustic atmospheres, areas to think, spaces to talk, ways to escape, modes of expression, and are enjoyed both alone and with others. As a moving space bounded by metal, glass, plastic and sometimes wood, it does not take a phenomenologist to point out that cars move us in more ways than one. Cocooned and private while wrapped in glass, cars are prime affective spaces for inhabitation. Of course, in addition to this positive view, daily experience for many includes anxiety, frustration, annoyance, and anger, which is understandable given lack of control (such as in heavy traffic), delays, accidents, and what in general is a high level of effort and attention, each of these potentially amplified for those whose living depends on driving. Notably, negative emotions such as anger are found to increase accidents (Zepf et al. 2019, 2020). Other emotions are more subtle, such as that unique experience of driving long distances late at night on empty highways (perhaps a mix of calm, introspection, freedom, and liminality [the sense of being in-between]). Carmakers and advertising agencies working on their behalf have long been sensitised to the affective nature of the car, with Volkswagen for example making a selling point of the emotion experienced when closing the car door of a Volkswagen Golf. Bucking usual aspirational car advertising, ads created by advertising agency DDB London, for example, depict several people in different situations presented with vehicles described as 'like a Golf', concluding with a woman in a car showroom visibly unimpressed by a salesman's assertion that a closing door sounds 'just like a Golf'. Then featuring a real Volkswagen Golf, the sound of the shutting door is reassuringly deep. The ad closes with 'Why drive something like a Golf when you can drive a Golf?' The developmental background is interesting, to illustrate designed affects. Neuromarketers studying 'sensory branding' find strong consumer interest in the sound of closing doors, and consumer belief they can distinguish between brands

(Lindstrom 2005), foremost in Japanese and American markets. This has led to the development of departments of sound engineers, product designers and psychologists to ensure that brands relay a sense of trust, safety, and luxury (McStay 2013).

It is from this affective, experiential, and communicative context that changes in automobility are emerging. Located within the study and practice of automotive user interfaces, driver state monitoring includes (1) real-time *safety* assessment of fatigue, driver distraction, intoxication, stress, anger and/or frustration; and (2) in-cabin *personalisation* through on-board assistants with voice sentiment capability, light adjustments, scent, seat vibrations (in tune to music), sounds, and heating levels. As will be developed, while driver state monitoring may involve outsourcing limited control to the vehicle; distraction, cognitive workload and emotion are better understood as automotive user interface topics for mostly *manual* cars (Ayoub et al. 2019). With drivers ultimately in control, a multitude of sensors may be deployed to gauge the body, affect, and emotion to generate holistic data about human behaviour. Measures potentially include heart rate variability, respiration, radar, voice, and touch-sensors, on sites such as the driving wheel and seats (including dedicated child seats). Signal types are from cameras (facial expressions); Controller Area Networks (CAN) Bus[2] (car speed, acceleration, and steering angle); wearables (heart rate, heart rate variability, respiration, electrodermal activity, skin temperature); steering wheel (electrodermal activity, skin temperature); and infotainment systems (touch interactions, voice interaction; adapted from Zepf et al. 2020).

Standalone cameras are currently most popular. These are less about good method or scientific need, but cost, complexity of installation and industrial expedience. For those looking to deploy in-cabin sensing there are trade-offs to be considered. A camera has limitations, but it is cheap and straightforward to install. Multimodal sensing through cameras and touch-based sensors is the ideal, but this array would involve cost and complexity. As developed below, the industry view is that a basic camera enabler is the place to begin, with view to increasing sensing modalities and services.

Yet, as with many of the examples of automated empathy in this book, there are historical examples of a nonoptical sort that preempt technologies, hopes, and worldviews of today. In the case of cars and purported in-cabin measuring of experience, this has origins in cars in the 1930s which was 20 years before introduction of the three-point seatbelt. The inventor Charles W. Darrow's (1934) 'Reflexohmeter' (introduced in Chapter 2) promised pocket-sised galvanic skin-response measurements outside of the laboratory. Applied to cars, the idea was that electrodes were applied to the hand and wires ran up the coat sleeve into a coat pocket, where the galvanometer is placed. Darrow asserts:

> The wider applicability of this stable pocket-sized galvanometer equipment is shown by a preliminary study of galvanic reactions while automobile driving. The powerful effect of any situation requiring the mobilization of the driver's energies, such as the slowing of a car in front, the attempt to pass another car on the road, the sudden appearance of a pedestrian in the road, or the sound of a policeman's whistle, is clearly demonstrated. Observations on the reactions of passengers to different situations are also of interest. (1934: 238)

In some ways little has changed in that cars are contained and relatively predictable spaces, with occupants mostly facing in one direction, making them advantageous for modern emotionology experiments. In addition to driver and passenger position (facing forward, except in the case of some infant seats and when reversing), driver behaviour is predictable in terms of human mechanical movements, gaze, and touchpoints (steering wheel, car seat, panel buttons, and gear stick). Objectives are also mostly defined, which means that in-cabin systems can focus on safety and enhancing experience.

Today there is widespread industrial interest in multimodal sensing (faces, voices, and bodies) to establish anthropometrics and awareness of qualitative human states such as drowsiness and emotions, and interaction between people in the car. Affectiva (who primarily work with facial coding of expressions) and Nuance (who are associated with conversational AI and voice analytics) for example teamed 'to deliver the industry's first interactive automotive assistant that understands drivers' and passengers' complex cognitive and emotional states from face and voice and adapts behaviour accordingly'.[3] Other affect-based start-ups point to 'redefinition' of in-cabin experience through biometrics, with Cerence speaking of the digitally enabled 'soul' of a car, also promising to redefine 'how a car should feel, respond, and learn on the go.' Others, such as Xperi, offer 'driver monitoring, occupant monitoring, iris identi-fication, advanced biometrics, and more'. A well-known start-up, it is no-table that Affectiva was bought by Smart Eye in 2021, a deal informed in part by merits of collaborating rather than competing in the automotive market for in-cabin sensing. This merging represents belief that the au-tomotive sector will be a successful sector for affect-based start-ups, and that this consolidation will grant recognition that they can compete with larger 'connected mobility' system suppliers to the car industry (such as Bosch, Ericsson, and Thales).

Driving Emotion: Safety

The 'autonomous car' is inherently strange because historically automobility is about human autonomy, based on freedom from reliance on animals (horses) and rail lines (collectively used carts travelling along fixed pathways at prescribed times). Today the expression 'autonomous car' has stuck, meaning the development of connected cars, grades of au-tonomous vehicles and vehicle ecosystems. It remains an open question of whether we will see fully autonomous cars on roads, but what is clearer is that travel infrastructure will feature cars with a level of autonomy. SAE International (2018) provides the definitive taxonomy for 'Levels of Driving Automation' for industry and policy. Scaled 0–5, the first set of levels (0–2) are defined by the driver still being in control over the vehicle, even though they may be assisted in some way (such as not having feet on the pedals). Key here is that the driver is supervising and maintaining the safety of the vehicle. The second set, 3–5, means that the driver is not driving, despite being in the driving seat. Levels 2 and 3 are of special interest because cars are entering the market that allow for collaborative driving (Level 2), where the car is handling parts of the driving, but the driver is still engaged. Likewise, Level 3 cars that allow the driver to disengage from the driving task in defined situations are also entering the market.

Use of automated empathy is not typically associated with fully autonomous driving (Levels 3–5), but lower levels. Cameras may detect activity type, emotional behaviour, and bodily behaviour, and take safety-related actions, such as briefly going into Level 2 autonomous mode until the driver is able to resume control. Automated empathy in cars should be seen in context of the wider industrial effort for adaptive systems and connected cars, meaning cars themselves that send and receive data from cloud-based services. That cars are part of road networks but not digital networks is a historic aberration, given long use of location tracking in smartphone apps and other systems. One could easily foresee for example car device IDs linked with Global Positioning System (GPS) data on road type, congestion, stopping, braking, turning, acceleration, and driver physiology to create safety-led cartographies of emotional geographies, potentially enabling drivers to prepare or amend their emotio-geographic journeys accordingly. Less speculative, automated empathy safety measures include sensors to detect usage of phones and in-cabin occupancy; states such as fatigue, drowsiness, distraction, intoxication, attention, and stress; affective states such as excitement and relaxation; and expressions of basic emotions, such as fear, anger, joy, sadness, contempt, disgust, and surprise, as introduced in Chapter 2. At a minimum this would involve prompts and notifications, 'You appear tired, why not take a break at [brand] service station off Junction 23 in approximately 21 minutes?' In-cabin profiling also raises the question of what happens in real-time if a driver is judged to be acting dangerously. Experts in human factors engineering and transport systems see the most controversial stage at Levels 2 and 3 of autonomous driving, because this is 'collaborative' where full autonomy is possible, systems may take over, and systems may pass back control to human drivers if there is a problem the car cannot resolve.

At this level, in the name of safety, such systems would not let drivers take back control if they are too emotionally aroused or at risk of being panicked by having vehicle control passed back to them, potentially going into minimum risk mode, and pulling over to the side of the road (McStay and Urquhart 2022). Such safety-led ideas raise many problems, especially involving *unnatural mediation* and *hyperreal* concerns, whether all people are sensed and treated equally, and [inadequacy] of emotion expression recognition. Moreover, even then the most beneficent approaches entail a deep paternalism that butts against ingrained cultures of driver self-determination.

The Whole Cabin Experience

Despite in-cabin space being relatively controlled from an experimental point of view as it involves mostly predictable tasks and activities, the messiness of human life features too. The car is a site for all sorts of experiences and activities. It is a place for conversation, being alone, refuge, calming pets, managing children, work calls, excitement, and being entertained. Human experience in this context is modulated by air conditioning (if even just windows up or down), ambient temperature control, cushioned and adjustable seating, and on-board communication and music systems. Indeed, on the latter, the coterminous nature of cars, privacy, intimacy, and entertainment reach back at least as far as the 1930s, as US auto manufacturers associated radio entertainment with individualised listening in automobiles and began to fit radios by default (Butch 2000). Today, interior experience is important for car makers. If in-cabin personalisation fails, this poses annoyance and a reputational risk for companies operating in a sector where branding is required to stand-out from other manufactures making functionally similar objects. Automotive user interfaces show interest in deepening and naturalising engagement with the car through greater attention to tactility, haptics (i.e., interested in touch and pressure), gestures, behaviour, expressions, and what are more sensational and feedback-oriented approaches to in-cabin experience.

The 'whole cabin experience' proposition presents challenges and risks, as well as potential gratification. Kia's 'R.E.A.D. System' for example claims to analyse a driver's emotional state in real-time by means of their facial expressions, heart rate and electrodermal activity—all in relation to baseline of driver behaviour. The system then tailors the interior environment, 'potentially altering conditions relating to the five senses within the cabin, creating a more joyful mobility experience' (Kia 2020). Similarly, Nuance's in-cabin assistant 'Dragon' (with Nuance being owned by Microsoft) reacts to the tone of a person's voice. If deemed joyful, the assistant will alter its language selection using more verbose words such as 'cool' and 'yep', whereas more neutral human commands will elicit more matter-of-fact responses from Dragon. Angry commands will reduce interaction to 'yes' and 'no'. Although one must be careful not to take marketing rhetoric at its word, especially given the clear flaws of hyperreal emotion profiling, one should not overlook the corporate money and expertise spent on human-factors engineering, occupant experience, need for competitor differentiation, and building of brand loyalty and aspiration through attention to affective factors of automobility.

Even without automated empathy, the act of driving 'is implicated in a deep context of affective and embodied relations between people, machines and spaces' where 'emotions and the senses play a key part' (Sheller 2004: 221). This spills into questions of affective approaches to economics, identity, culture and experience-based understanding of automotive design, production, and marketing. Sheller speaks of libidinal engagement where, 'Touching the metal bodywork, fingering the upholstery, caressing its curves, and miming driving 'with all the body' suggests the conjoining of human and machinic bodies.' One can debate whether luxury and libidinal (as per Freud and his principles of desire) equate to the same thing, but Sheller is right to focus on touch, sensation, and affective dimensions of a more fundamental nature than symbolic and image-based capitalism. To the affective context of car production automated empathy adds a relational dimension enabled by tracking of kinaesthetic, pressure, gesture, gaze, tone, and in-cabin physiological behaviour. Uniting the two cultures (humanities and science), this is an affective physics through empirics and preemption of experience. Indeed, under the auspice of monetisation and new services, mission creep with data is occurring before safety driver-monitoring has reached the roads.

REGULATION: THE VISION ZERO TRADE-OFF

Of all the uses cases of automated empathy discussed in this book, cars are especially interesting because automated empathy has the backing of policymakers seeking to make driving safer. At an international level is the United Nations initiative 'Decade of Action for Road Safety (2021–2030)' that operates under resolution A/RES/74/299 (WHO 2022) with the UN also naming road safety in its Sustainable Development Goals (SDGs, see 3.6 and 11.2). In Europe, as of mid-2022 in the European Union (EU) all new cars in the EU market will be equipped with advanced safety systems due to the coming into force of the EU Vehicle Safety Regulation 2018/0145 (European Parliament 2018). This entails numerous factors of interest to the study of automated empathy, potentially including in-cabin alcohol testing (where a sample of alcohol-free breath is required before the vehicle will turn on), driver drowsiness and attention warning systems, advanced driver distraction warning systems, and event data recorders. The EU's Vision Zero initiative is based on the Swedish approach to road safety where no loss of life is acceptable. Statistically, this is interesting. Globally 1.35 million people die per year of road death (the world population is around 7.6 billion). The European region has the lowest rate, at 9.3 deaths per 100,000 population, with the African region having the highest road traffic fatality rate, at 26.6 deaths (WHO 2016).

The Vision Zero initiative recognises that people make mistakes, and that traffic needs to flow, so urges mitigation measures to reach Vision Zero (TRIMIS 2021). The working document to reach Vision Zero by 2050 is the EU Road Safety Policy Framework 2021–2030, recognising that reducing road death is multifaceted (involving non-car and technology solutions) technological advances are prioritised, 'foremost in connectivity and auto-mation' (European Commission 2020: 7), which will in future create new road safety opportunities by reducing the role of human errors and 'taking the physics of human vulnerability into account' (Ibid.: 11). This is a choice in that urban policy has chosen to focus on technological solutions and surveillance (systems of seeing and control), rather than means of re-ducing car traffic. Vehicle safety measures include passive safety features such as safety belts, airbags, and crashworthiness of vehicles, but also active safety features, such as Advanced Emergency Braking, Intelligent Speed Assistance, Stability Control, and Lane Departure Warning that may prevent accidents from happening altogether.

The EU Road Safety Policy Framework 2021–2030 document also encourages fitting of 'state-of-the-art advanced safety technologies' in ref-erence to the European New Car Assessment Programme (Euro NCAP). NCAP industry-oriented programmes will be discussed below, but for now it is simply worth noting that NCAP is an industry friendly global programme with local presences that assesses and rates cars for safety. From the point of view of automated empathy and critical questions about attempts to gauge human interiority, there are three key aspects to the EU Road Safety Policy Framework 2021–2030 document: (1) it states that EU Member States should be able to access in-vehicle data to determine lia-bility in case of an accident; (2) recommendation of consideration on the collection of anonymised data about car safety performance; (3) broad interest of the Framework document in 'more complex human-machine interfaces' (2019: 18). The first point raises questions about privacy and transparency of accountability in cases of incidents; the second about the value of assurance of anonymisation; and the third about the nature of human-machine interfaces.

On the second point about anonymisation, the EU General Safety Regulation (2019/2144) for cars (the aforementioned law regarding new cars on the EU market featuring advanced safety systems) mandates for distraction detection, particularly for warning of driver drowsiness or dis-traction. Recital 10 therein states 'Any such safety systems should func-tion without the use of any kind of biometric information of drivers or passengers, including facial recognition.' Given that sensing and data processing of drowsiness or distraction must occur somehow, the gist is acceptability of the premise, but that human-state data should not identify a person, because biometrics are considered as a special category of personal data. Yet, if the data does not fulfil the criteria of personal data (the ability to 'single-out' a person in some way), then rules about biometrics do not apply, because as the law understands biometrics here, they are contingent on identification. Rather, systems may use edge and fog computing where data is processed and discarded in the car or closely related devices such as a smartphone (the personal data exhaust). This is made explicit in Regulation (2019/2144), with Recital 14 being surprisingly explicit, stating that personal data, such as information about the driver's drowsiness and attention or the driver's distraction, should be carried out in accordance with the EU GDPR and that 'Event data recorders should operate on a closed-loop system, in which the data stored is overwritten, and which does not allow the vehicle or holder to be identified.' Furthermore, systems should not 'continuously record nor retain any data other than what is necessary in relation to the purposes for which they were collected or otherwise processed within the closed-loop system' (Ibid.). Later, Article 6(3) of the Regulation further states that not only should such systems be 'designed in such a way that those systems do not continuously record nor retain any data other than what is necessary in relation to the purposes for which they were collected or otherwise processed within the closed-loop system', but 'data shall not be accessible or made available to third parties at any time and shall be immediately deleted after processing'.

The significance of both Vision Zero and EU General Safety Regulation (2019/2144) for automated empathy, and in-cabin technical feeling-into of fatigue and drowsiness, is that all forms of sensing and surveillance are permissible if no data is retained that identifies a person. This is much more than legal technicality; it has social significance. It is the proposition that social harm through data is reducible to privacy; that privacy is reducible to the principle of whether a person is identified; that temporality (near immediate deletion of data) is a solution to balancing benefits with harms); and that if these can be resolved, then all forms of sensing and actuation of processes are thereafter fair game. Non-identification as the royal route to environments based on automated empathy is one that recurs in other quite different use cases (McStay 2016, 2018).

Proxy Data Mosaics

The European Data Protection Board is the body tasked with ensuring that EU member states properly apply the General Data Protection Regulation (GDPR). In its guidelines on connected cars (EDPB 2020) it raises numerous issues. First is the high-level recognition that the EU legal framework for connected vehicles is based on the General Data Protection Regulation (2016/679) and the 'ePrivacy' directive (2002/58/EC, as revised by 2009/136/EC). The connected cars guidelines state that indirectly identifiable data includes the details of journeys made, the vehicle usage data (e.g., data relating to driving style, or the distance covered), or the vehicle's technical data (e.g., data relating to the wear and tear on vehicle parts). Notably too, even though (for example) distance covered cannot identify a person, that such data may be cross-referenced with other data (especially the vehicle identification number) means it *can* be related to a natural person to build a mosaic of information that acts as a proxy for directly observing drive behaviour.

Although the biometric aspect of automated empathy is likely to capture critical attention over vehicle system data, both are topics for automated empathy due to the use of proxy information to gauge interiority, for example through analysis of a driver's steering behaviour during one or more trips based on information from changes in the steering-angle sensor in a steering wheel. Small as well as larger changes for example signal lapsing concentration, especially when combined with contextual data on the frequency of movements, length of a trip, use of turn signals and time of day. Consequently, lending weight to the problematic nature of collecting any data about telematics (technology to monitor information relating to vehicles), even innocuous information potentially acts as pieces of a mosaic to profile people. Consequently, the Board (following

GDPR) states that *any* data that can be associated with a natural person is within scope of regulation.

The policy goal then is privacy-by-design solutions which enable safety measures but do not fall foul of data protection rules. Practically, this might entail operating the car's systems through a driver's smartphone and not sharing insights with car manufactures or suppliers who have networked hardware in the vehicle. With this meaning that the user has sole access to the data, the EDPB thus suggest that 'Local data processing should be considered by car manufacturers and service providers, whenever possible' (EDPB 2020: §72). *If* data needs to leave the car, the EDPB recommend both anonymising processes that sever links between a person and dataset (i.e., removal of identifying attributes) to avoid coming under the jurisdiction of European data protection rules. In theory then the presence of automated empathy in cars is about impossibility of data mosaics due to rules on sharing of telematics about vehicle usage, almost immediate deletion of in-cabin occupant data about human states (such as emotion or fatigue), and belief that privacy harms can be managed through rules on identification and data retention time. As developed below, compliance is already looking problematic, arguably due to the appeal of having access to these insights, or difficulty in complying. This in large part is due to the desire for connectivity and the impossibility thereof of ensuring privacy. For example, Bosch (the German engineering and technology company) promote their new approach to automobility, claiming that 'Users, vehicles, and surroundings are seamlessly connected, making driving more enjoyable and providing a personalized mobility experience.' The rhetoric is noteworthy, stating that in-cabin profiling is the vehicle's ability to 'keep a protective eye on their occupants' and that the interior monitoring system ('Guardian angel') detects drowsiness, distraction, unsafe seating position, warns inattentive drivers, recommends a break if drivers are perceived to be tired, and may reduce vehicle speed (subject to the automaker's wishes and law) (Automotive World 2021). The risk is clear in that the EPDB seeks to reduce both the lifetime of personal and restrict its movement to a driver's devices, while marketing-speak of manufacturers such as Bosch promote a much more expansive use of occupant data, one involving surroundings, connectivity, and likely third-party usage.

Voluntary Programmes: Euro NCAP

To understand both impetus and constraints regarding automated empathy in cars, one should appreciate the role of the New Car Assessment Programme (NCAP). This is a published set of standards originated in the US. Expanded to Europe as the European New Car Assessment Programme (Euro NCAP) it provides European consumers with information regarding the safety of passenger vehicles, publishes safety reports on new cars, and awards 'star ratings' based on the performance of the vehicles. Historically, this is in relation to all likely collisions that a car and its passengers might be involved in. In addition to collision standards, Euro NCAP Advanced was set up as a reward system launched in 2010 for advanced safety technologies. Since 2020, Euro NCAP requires *driver monitoring* for five-star vehicle ratings. This is 'to mitigate the very significant problems of driver distraction and impairment through alcohol, fatigue, etc.' (Euro NCAP 2018: 2). As with systems outlined above, Euro NCAP also promotes systems that monitors drivers, detects problems, warns drivers, and then potentially limits speed. This is said to take the form of camera-based techniques that measure head and eye movement (and facial expressions), but other options include heart rate detected by the car steering wheel.

I joined several panel webinars hosted by Affectiva in 2020 about the automotive sector. These proved highly informative because they involved input from technologists, industrialists, and governance insights from Euro NCAP.[4] Topics covered included how Euro NCAP is influencing the development of driver monitoring system, safety features, and in-cabin sensing beyond safety features. Although each were hosted by a US-based company, featuring panellists from US companies and universities, it is notable that it was about *Euro* NCAP rather than other jurisdictions. Across the webinars it became clear that solutions and performance indicators on how best to comply with Euro NCAP regulations was being left to the industry itself, although Richard Schram (Technical Director for Euro NCAP) stated that any system deployed 'needs to work in real life' given the lifecycle of cars and need for reliable safety provisions. Notably Schram was not keen on regulation, preferring that the market (car manufactures and second tier suppliers such as Affectiva) set workable standards (albeit within the confines of the law).

Notable given this book's critical alertness to questions of *unnatural mediation*, Schram highlighted that any proposed systems or in-cabin sensing of human states and emotions should recognise cultural difference and diversity of languages, although this was not explored in depth (leaving open questions of overall ethnocentric differences in expressiveness, problems for individual drivers deviation from an intra-regional emoting threshold, along with other concerns associated with unnatural mediation). Representatives from industry did not defer from Schram's concerns and it might be noted that Caroline Chung of Veoneer and Rana Kaliouby of Affectiva have non-White/non-US backgrounds. This complicates the 'white tech bro' narrative, especially given Kaliouby's Arab-Muslim Egyptian background, and Affectiva's status as the preeminent start-up in emotion sensing.

In relation to accuracy, regulatory concern about driver profile mosaics and connectivity faces another challenge. Whereas Bosch was cited above claiming need for enhanced connectivity on grounds of safety, the agreed suggestion by those involved in the emotion recognition industry, and Euro NCAP, is to connect multiple human-state signals to ensure accuracy. All believed that effectiveness would be increased by multi-signal reading of drivers (including gaze direction, face expressions, voice, body movements, and detection of advanced driver states to capture onset of drowsiness before a driver has 'microsleeps'). Just as Chapter 2 cautioned against bias against marginalised groups through unnatural mediation as the prime prism through which to criticise hyperreal emotion/affect profiling, the same applies here: argument-by-accuracy may amplify the activity that is of concern. The risk for critics is the noise around accuracy inadvertently misdirects from harms caused by industrial solutions, that is, more, not less, surveillance through holistic assessment of driver biometrics. Effectiveness, accusation of pseudoscience, and accuracy are not the primary issues, although they are important secondary concerns. Central is the relative merits of use of automated empathy, if at all.

Sensing Not Storing?

Use of telematics to monitor driving behaviour and recommendations of coffee when fatigue is detected may be deemed a good thing. Indeed, use of cameras and biometrics to detect eye-flutter and change in face expressions may also be considered entirely acceptable surveillance in the name of safety. The stance of hybrid ethics outlined in Chapter 4 keeps the book open to new uses and applications of technologies, even when operating with much-criticised technologies, such as biometrics. A key question is whether safety is all there is to it: a closed loop, as regulators suggest, where data is immediately discarded, avoiding chance that biometric and telemetric data might be used in unexpected ways. All present on the Affectiva hosted webinar recognised the sensitivity of in-cabin sensing, noting privacy concerns, user acceptance, need to generate trust invisibility to users, and the need for users not to switch systems off. Yet, the panel posited that if people are happy to give data to Facebook, why not for their own in-cabin safety? (It is of course debatable that people are happy to give data to Facebook/Meta.) Persuasion techniques and 'socialisation' were mentioned, such as marketing these technologies based on child as well as driver safety, but also the panel's belief that posts such as 'My car has saved my life' will emerge on Facebook and help create adoption of in-car sensing of bodies. Echoing EU legislation, Schram was keen to assure webinar delegates who asked questions about privacy that the car is *sensing rather than storing*. Yet, a throw-away comment Schram (11/02/21) is attitudinally telling, where he remarked 'Don't rock it for me because you don't like it' and that 'only a small percentage of people complain about privacy'.

In some leading use cases there is no debate about whether a person is having data collected about them. Bosch for example unambiguously state that 'Facial recognition makes it possible for the system to identify the driver with absolute certainty' and that 'Based on the stored driver profile' the system will set personal in-cabin preferences (comfort, media, temperature, and mirror positions; Bosch 2022). Other approaches however are based on sensing not storing. Although sense data may not be stored, it may inform other systems. For example, if a system senses the beginning of fatigue it might suggest upbeat music. The biometric data (e.g., face and heart rate) may be discarded instantly but the information (that a person may have become tired) will live on by other proxy-based means (such as a recorded uptick in music recommendation or cooling of in-cabin temperature). As these systems are deployed, privacy is clearly going to be a sticking point because there is a clear scope for contradiction between claims of 'sensing not storage' versus the strict EU rules on data processing, especially if consent is going to relied upon as the basis by which to process personal data.

Such systems will also fall foul of 'transient data' concerns. Here, systems that promise sensing not storage *do* temporarily store personal data by means of a transient copy temporarily kept in the camera system memory (George et al. 2019). Although this is only for a fraction of a second, so to an extent a technicality, European data protection law itself does not stipulate a minimum length of time that personal data should exist before being subject to the law. Nevertheless, this raises a question that applies to many use cases of automated empathy: is making human physiology and emotional life 'machine-readable' justifiable *if* the data held and used by a system does not identify a person (beyond the fraction of a second identifying data is held)? Might it exclude application of the GDPR and similarly founded data protection legislation altogether (also see McStay, 2016, 2018, McStay and Urquhart 2019)? George et al. (2019: 286) cite one system used in out-of-home advertising that watches people back where: (1) the original pictures are neither stored permanently nor processed in a cloud; (2) a transient copy is temporarily kept in the camera system memory; (3) the camera software detects the characteristics (metadata) of the person and generates a hash value of the metadata; (4) this process takes fractions of a second; (5) the generated metadata in George et al.'s adtech example is used to choose an advertisement and is deleted after 150 milliseconds. The justification is that patterns and valuable insights may be derived without impacting on individuals. Legally this gets too complex for our purposes (law scholars should see George et al. for discussion of Articles 11 and 15–20 of the GDPR, identification and terms of exemption), but their conclusion is a shift in jurisprudence from technical questions of input to effects. This would allow for more holistic assessment of data processing, such as scope through discrimination and treating groups of people differently, as is the case with nonpersonally identifying data. It is eminently sensible in that (to cite George et al.'s example) citizen benefits of advertising that watch back for emotion expression reactions are few, if any. While there is an argument that car safety is a different proposition, and that transient data is a good data minimisation principle, what of the possibility (or likelihood) of spilling into entertainment and soon thereafter advertising?

Indeed, data processing designed for safety and in-cabin personalisation may leak into entertainment systems, therefore opening the way to advertising and marketing based on biometrics soon thereafter. Given that data about emotion is already thought by citizens (certainly in the UK) to be intrinsically valuable (McStay 2018), the details of what, where and how data is processed is going to require careful explanation in marketing as well as opt-in consent statements. Yet, in the webinars Andy Zellman from Affectiva (11/02/21) remarked that 'We give it [data privacy and information about purchase habits] away on a daily basis,' raising questions about sincerity to protections from unwanted marketing techniques, the degree of control a driver (and other occupants) would have over in-cabin personalisation. Although across the three webinars opt-*in* approaches were stated as the preferred method of gaining consent, as numerous studies on consent show, the voluntariness of opt-in is of course dubious in practice, subject to 'dark design' practices that mitigate against free choice.

Pooling Empathy: A Consortium of Emotions

An idea raised on one of the Affectiva webinars is that of a consortium of standards regarding data about emotions and affects, involving a shared industry resource that would define industry standards for when a named emotion (such as anger) is said to have taken place, or the criteria for when a person may be labelled fatigued or distracted. Currently, there are no agreed standards among car manufacturers, those supplying them with systems, or those interested in this data, regarding standards for the nature of emotion and affects. This contemporary interest in an industrial and academic pooling of data to create a common set of standards regarding emotion expression, thresholds (such as fatigue and when microsleeps begin) and other human-state measures is important. From the point of view of industry, it would create a set of identifiable standards by which applications would function, but for critical onlookers it would make transparent industrial understandings of psycho-physiological states and emotion. This might for example assist product safety assessment under the EU's proposed AI Act that requires certification for High-Risk AI Systems such as emotion recognition systems used in safety equipment.

Such standards would have to not only answer the problem of regional averages but, given the sensitivity of decisions to be made, also the problem of what intra-societal normality is. While a societal mean average of features or emotional behaviour might be generated, all people will deviate from the average in some regards. The upshot of this is that an ideal is something reality is measured against. This observation has origins in the 'moral statistician' Adolphe Quetelet (1842) who pointed out that all members of a society will be a deviant because few, if anyone, is perfectly average. Quetelet's conclusion is that the individual person who is compared to the ideal is synonymous with error. This lends to the hyperreal and unnatural mediation arguments introduced in Chapter 2 because all people will stray to some degree from the societal mean average of features and emotional behaviour. It is hyperreal in that the ideal has controlling presence, but it is an artifice that people would be required to live-up to.

Such a set of technical standards would need to be verified, published, and subject to scrutiny. It would also need to be dynamic enough to operate in all regions where cars are sold, thus providing a litmus test for all question and misgivings about the difficulties of creating global standards for behaviours that have profoundly contextual characteristics. Although this book sees no merit in use of systems based on hyperreal emotion, other parts of automated empathy for safety (such as gauging of fatigue and distraction) are arguably more worthy. Standards for human-state measures would however need to contend with the problem of Quetelet-based deviance from ideal typification, and potentially liability and insurance matters that flow from these in case of accidents. Other practicalities take the form of incentives in that while all in the industry sees the advantage of a consortium of standards for human-state measures, there is an investment imbalance between car makers and their suppliers of parts and systems; and those who create automated empathy systems that may be used in cars but are not designed specifically for them. Given problems of unequal investment into such a consortium and who stand to benefit most, Euro NCAP's Schram suggests that the consortium should be public, collaborative, and based on shared data, which invites potential for academic and regulatory scrutiny of the effectiveness of in-cabin sensing. Scope for public scrutiny by independent experts of how and whether systems work is a minimum requirement.

THIRD PARTY INSURANCE

Any suggestion that vehicle and biometric data will not go further than the car itself is not at all assured, especially when one considers ambition for 'connected cars' involving 'connecting users, vehicles, and services over the internet' and that 'drivers also get a fascinating experience and ultimately have more fun behind the wheel' (Bosch 2022a). There will be many Internet-connected interested third parties, including media and entertainment services, advertisers, service stations, breakdown services, and insurers. The insurance industry will be interested in personal and/or aggregate understanding of both car and driver behaviour, likely offering lower premiums to those who accept in-cabin profiling.

Insurance is a sector oriented to data and information to gauge risk and the likelihood of something happening. Profit is generated by how well it understands risks that it insures against. The promise by the insurance industry will be clear: 'Give us more data so we can better determine risk and we in return will reward those drivers who drive carefully.' One major problem with this is that insurance is innately collective, being based on pooling of risks. In context of increasingly connected objects (such as cars), profusion of data, and computers tasked with identifying and learning about behaviour, insurance is shifting towards individualised risk. Exploring big data and insurance, Barry and Charpentier (2020) cite O'Malley (1996) to distinguish between 'socialized actuarialism' (that aims for large pooling of shared risk) and 'privatized actuarialism' (that puts responsibility for managing risk on the individual). Whereas historically risk is something that has been shared between insurance users, aiming to ensure that all may receive cover, insurance practice is becoming about predictions of individuals.

As it stands, above-discussed EU law on transport gives the impression that safety systems will not in principle store data about people. Yet, as also noted, if cars are to offer tailored experiences through telemetry, then while biometric data may not be stored, it can certainly be inferred. One can foresee pressure from road safety groups for this to be recorded, for instance in situations where a driver was repeatedly warned to take a break but chose to ignore it, causing an accident. Critical awareness is required because risk profiling in the insurance sector has long used proxy information to help with pricing risk. Marital status, age, and postcode for example are proxies of risk when considering fraudulent claims for whiplash, which is commonly caused by rear-end car accidents (McStay and Minty 2019). Insurers are naturally interested in ability to track eye and hand movements in both fully manual contexts, but also assisted driving (Levels 2 and 3 of the SAE scale) when a car hands back control to a driver. On emotion, Allstate, (a US insurer) wants to use data from the telematics of black boxes fitted to the cars it insures to identify drivers who show symptoms of stress, which would then be used not only for motor underwriting, but life and health underwriting too. This approach to personalised insurance is based on the actuarial principle of adverse selection (McStay and Minty 2019), which says that it would be overall unfair on customers of an insurer not to charge an individual a higher premium when they are identified as a higher risk, relatively to the premium it charges for a lower risk. For automated empathy, scope for enhanced understanding of in-cabin behaviour will put pressure on insurers to make use of empathy and other biometric data, lending impetus to the suggestion that keeping data in a car that is built to be connected is an unrealistic proposition. However, another outcome regarding insurance is possible: if regulators see problem with third-party usage of car data, car manufacturers themselves may offer the insurance, something that regulators may find appealing due to this being a first-party relationship (McStay and Urquhart 2022).

AUTO-VULNERABILITY

This chapter's interest in cars continues by exploring the potential for affective states *in cars themselves*, not their occupants, involving what McDuff and Kapoor (2019) posit as 'visceral machines'. With Daniel McDuff and Ashish Kapoor both being based at Microsoft Research, this is not idle fancy. Visceral cars are conceived with a greater degree of autonomy than discussed so far, operating at levels 3–5 of the SAE International scales of driving automation, meaning that the driver is not driving, despite potentially being in the driving seat. The visceral proposition is due to the proposal that cars may function by simulating aspects of the human autonomic nervous system. For McDuff and Kapoor, equipping a car with ability to respond to threat is a good way to manage danger, risk, and increase overall road safety. Based on psychological interest in the role of affect in making decisions, this involves threat assessment and immediate action, but also planning, decision-making, and learning.

Modelling computer systems on the nervous system is not new. In 2001 IBM launched 'autonomic computing' that was modelled on the auto-nomic nervous system (IBM 2005; Hildebrandt 2011). IBM's stated goal involves computers that would self-configure, self-heal, self-optimise, and self-protect. In academic work, Man and Damasio (2019) note Di Paolo's (2003) favouring of organism-level logics, susceptibility to en-vironmental cues based on abstract internal variables (Parisi 2004) and explicit references to homeostasis (Di Paolo 2000). Yet, the approach by McDuff and Kapoor (2019) is different, with systems being trained through supervised learning of people and their responses to danger. The system does not mimic biological processes, but it is based on insights derived from human peripheral blood volume pulse, which measures the volume of blood that passes through tissues in a localised area with each pulse or beat of the heart. (In a domestic context, this can be measured with a clip-on finger oximeter.) Changes in rapidity are argued to indicate situations that correlate with high arousal of a person's nervous system. Human physiological responses to danger were then used to train a neural network system that can predict, behave as a nervous system would, and learn to cope with new situations by means of reinforcement and rewards.

With affective states in cars themselves, the visceral safety-oriented system would (in theory) help cars appraise situations to better plan, react, and cope with changing environmental conditions. Broadly aligning with Nussbaum's (2001) appraisal and cognitive-evaluative account that connects emotion (and affects) with the perceived importance of people and objects, such a system would help with judging, perceiving value and ethical reasoning; the artificial nervous system would (in theory) help provide an affect-oriented salience for decision-making systems; it would help the car and occupants 'flourish'; and it would create salience of ex-ternal objects (and/or people) in relation the car's own scheme of goals. Notably, this is a subjectivist approach, based on mapping events things to the subject's own goals, primacy and to need to thrive.

McDuff and Kapoor (2019) note that when human drivers are traveling in a car at high speed, they will likely experience a heightened state of arousal. This response is automatic, involving heightened awareness to a range of factors, especially outside of the car (such as other traffic, position, road furniture, cyclists, and pedestrians). Their argument is that as a person's sympathetic nervous system guides emotional reaction and decision-making, then a form of this enabled in a car might help it avoid and cope with dangerous situations. This is because the sympathetic nervous system in people and animals provides ongoing feedback in relation to environments, objects, people, and other creatures. Alertness would 'indicate adverse and risky situations much before the actual end-of-episode event (e.g., an accident) and even if the event never occurs' (2019: 3). Like a person, alertness and arousal of the nervous system would vary in relation to situations perceived.

Such a system would not just scan, but also learn by means of reward mechanisms. The first reward mechanism is extrinsic and the second intrinsic. Extrinsic entails scoring and rewarding of behaviours that are task specific, meaning rewarding of autonomous behaviour that contribute to objectives of travelling large distances safely, being punished for collisions, and sub-awards for negotiation of features within journeys. Intrinsic entails rewarding of actions that lead to states that reduce stress and anxiety (meaning the car itself 'feels' safer). This calm-focused approach is a preservationist, rather than mentalistic, approach to ethics-by-design. At the time of writing this is early-stage research, yet the visceral and affective suggestions have intuitive appeal to automation of danger management. They also raise robo-ethics questions, especially regarding the ethical problems of technologies being programmed to 'feel' negative states such as anxiety (Wallach and Allen 2008; Gunkel 2017).

Visceral and auto-vulnerable cars are inextricable from ideas core to the origins of cybernetics in the 1940s, that involve goal-orientation, preservation, regulation, homeostasis, and behaviour. The goal of the visceral and auto-vulnerable autonomous car is to minimise the physiological arousal response, in effect keeping the car calm and nonstressed. It is also useful to recollect cybernetic principles sought to explain biology, as well as new ways of thinking about technological possibilities. As the science of feedback it is the study of circular and causal relationships between a system (an animal, person, machine) and its environment (Weiner 1985

[1948]). Thus, when the language of biology is employed today in relation to technology, the paradigmatic logics and principles are not new. Indeed, more recently embodied AI is explicitly based on approaches to machine intelligence, where intelligence emerges through interaction of a system with an environment and because of sensorimotor activity (Pfeifer and Iida 2004).

Applying embodiment to the rights and wrongs of ethics, Wallach and Allen echo emphasis on contextual and environmental factors in their account of artificial moral agents, in that the problem is not how to programme them with abstract theoretical knowledge, but 'of how to embody the right tendencies to react in the world' (2008: 77). This inductive approach to automated machine ethics is an attempt to by-pass mentalistic decision-making by taking recourse to programmed instinct, which also begins to address the critique of AI from Dreyfus and Dreyfus (1990: 331) that any conception of intelligence involves needs, desires, emotions, a human-like body, abilities and, most relevant here, vulnerability to injury. Leaving aside for one moment critical questions of how human values are programmed into systems, if one adopts a view of ethics based on bodies and flourishing (Lakoff and Johnson 1999) these 'right tendencies' are deeply biosemiotic in character. Thus, McDuff and Kapoor's (2019) visceral machines are not simply about safety and preservation, but instead the adequacy of visceral ethics, which is an embodied approach to ethics. Grounded in the body and experience, for philosophers of neuroscience bodies and biosemiotic flourishing are at the heart of moral concepts, which are 'inextricably tied to our embodied experience of well-being: health, strength, wealth, purity, control, nurturance, empathy, and so forth' (Lakoff and Johnson 1999: 331). Their argument is that to cut through the diverse and some-times opposing normative stances on ethics, an approach that starts from the body itself is fruitful. This is different from inductive 'organic ethics' in AI research (Wallach and Allen 2008: 99) in that a bottom-up approach to AI ethics proposes that systems may learn rights, wrongs, and human morality from observation regarding the acceptability [or not] of an action. The visceral and embodied approach is different, being affective and less cognitive in nature.

Yet, building ethics and safety into cars through the equivalent of feelings in a living organism *is* novel. The key difference is injection of vulnerability. For occupants, a car that is not anxious or behaving erratically is likely to make for a pleasant ride. Yet, in an autonomous context, biosemiotic reactions have social significance, because they impact on people and surrounding objects. Thus, whose goals does an autovulnerable visceral car serve? Certainly, a car that takes measures to keep itself calm will be of benefit to other people and things in its environment. However, being based on homeostasis (the stable and optimal conditions for a living system), survival, and self-preservation of the car's ideal affective state, it is in theory likely to prioritise itself and its occupants over elements (or people) within its environment.

McDuff and Kapoor's (2019) approach is only loosely biomimetic in that while it draws on learning about the affective states of human drivers, McDuff and Kapoor do not seek to directly replicate a person's sympathetic nervous system. Others approach the question of affect in technologies more literally, with Man and Damasio (2019) grounding their arguments in the idea that a machine can behave as a natural organism. Also discussing feeling, homeostasis, vulnerability, and risk to self, they draw attention to advances of soft materials that simulate the soft tissue of people and animals. Applied to cars, this takes us into strange territory, one that not only features simulation of affective states based on vision and electronic sensors that record disturbances, but vulnerability in the car materials themselves, based on the idea that if all cars were more vulnerable, such as exteriors of softer materials rather than metal, all would drive more safely. Unlikely perhaps, but given that the metal of dumb cars typically fares better than sensitive flesh of people and animals, if cars were to feel vulnerability (or less controversially, able to simulate the process of feeling vulnerability), this might serve as a good foundation for road behaviour and reduction of human road death, as per the UN Sustainable Development Goals focused on road safety (SDGs 3.6 and

11.2), the EU's Vehicle Safety Regulation (2018/0145), the EU's Vision Zero strategy, the EU Road Safety Policy Framework 2021–2030, and NCAP industry-oriented programmes.

Affect and the Trolley Problem

Auto-vulnerability provides a novel perspective on the infamous Trolley problem, now associated with AI decision-making and autonomous cars. This is the thought problem in which an onlooker has the choice to save five people in danger of being hit by a trolley, by diverting the trolley to kill one person. With a 'trolley' being known elsewhere as a 'tram', both are defined by running on track laid on public urban streets.

The nub of the problem is that a driver (human or otherwise) faces a conflict of negative duties since the driver should avoid injuring either five people or one person. With the vehicle originally being a tram and the problem created by Foot (1967) to explore intentions, decision-making, and 'double' effects of decisions in relation to abortion, it is a longstanding conundrum in moral philosophy where utilitarianism and deontological ethics clash over the dilemma of whether to intentionally end one life to spare numerous lives. This was adapted and expanded by Judith Jarvis Thomson to explore rights, medical ethics, organ transplants, surgeon decision-making, characteristics of the five and the one, and whether killing and letting die are the same thing (Thomson 1985). Thomson's take is based on a passing by-stander who sees the problem and might intervene by switching the rails a trolley is careering on (the difference is that a driver must decide, but a bystander even with a switch may still do nothing).

Applied to cars, one can see appeal. If an autonomous car is heading down the road and is faced with an unavoidable decision of either killing five people, or swerving to kill one—what it should it do? The principle of utility is clear on what it would advise, but other views struggle. A well-rehearsed problem, there are many approaches to the question. Some answer it and others reframe the problem. Although utility (saving the five) has appeal, as Thomson (1985) and Sandel (2009) suggest, if the situation is tweaked by restating that an onlooking person on a bridge can save the five by pushing a person off a bridge to stop the speeding vehicle, this becomes more problematic. Intuitively not right, such a scenario fails to recognise and respect individual rights. Enlarged, is this moral economy suitable foundation for a society that anyone would want to live in? The problem here is not the ratio (1–5), it is what society would be if established on the foundation of calculus. The nature of the cost is that it contravenes the Kantian principle that identifies people as ends rather than means, also avoiding granular but necessary human considerations of context, circumstance, alternatives, obligation, permission, infringement, motive, and rights. To continue the thought experiment for auto-vulnerable cars, what if the car had something to lose? This might involve extrinsic costs of having failed to protect itself, and intrinsic trauma of having deviated so far from the equilibrium of a calm affective state. The self-interested car would perceive danger and stop to protect itself, and others by default. By default, it would refuse moral calculus, normative choices, or judgement by virtue. A final twist on the auto-vulnerability angle on the Trolley problem is need for distrust in the seeming naturalness of biosemiotic solutions to safety. Problems of unnatural mediation are present, with one expert industry figure interviewed by McStay and Urquhart (2022) observing disproportionate social representation in training data for computer vision in cars. As part of the review team of an autonomous car failure, she found that due to not being trained with enough examples of women, the car mis-classified the woman as an object. Indeed, risk is amplified in an intersectional context, with facial analysis being disproportionately worse with women of colour than other groups (Muthukumar et al. 2018) and trans-people (Keyes 2018). The problem is hybrid, exceeding utilitarianism and deontology, involving human entanglement in things, in this case involving machine intentionality (accounted for in Chapter 4) and what exists for automated technologies. Despite the biomimetic and biosemiotic impetus that motivates ideas such as auto-vulnerable cars, traditional questions of how people are represented in computational and mediated understanding remain a foremost concern.

CONCLUSION

Examining the automotive sector, this chapter was split into two sections. The first assessed automated empathy as it applies to manual cars and in-cabin sensing of vehicle occupants. The second considered industrial interest in building autonomous cars that simulate feelings of vulnerability. For occupants, cars are important sites of feeling and decision-making. A panicked decision may lead to minor or major crashes, as might affective states such as sleepiness and fatigue. There are then safety incentives in use of systems that sense and react to occupants, as outlined in as variety of European policy initiatives. As the chapter found, systems designed for safety are prone to mission creep, where analytics designed for one purpose are used for other reasons. Key is that if such systems are to be used, safety and affect-based in-cabin personalisation must be kept entirely separate. Indeed, the hybrid, post-phenomenological and grounded stance of this book leaves it open to trialling automated empathy in cars explicitly for safety purposes. Maybe the ads for Volvo described at the top of this chapter have a point regarding changing norms, in that *if* data is immediately discarded and irretrievable, and human-state measurement demonstrably saves lives, such systems may be ethically sound. Yet, internal industrial attitudes that are suspicious of citizen preferences for data privacy should be guarded against. This requires the firmest of upfront regulation that deals carefully with the 'sensing not storage' claims, especially if personal data is found to be used, however transient it may be.

The latter section of the chapter noted industrial interest in cars that function by means of what is akin to a human autonomic nervous system, trained by data from people and their responses. There is novelty in this but also long roots in cybernetics and maintenance of equilibrium of a system, where a car seeks to reach a destination with minimum excitement of its systems. Despite the eye-catching title of 'visceral machines'

(McDuff and Kapoor 2019) and its linguistic association with human intuition and gut-feeling, affect-based auto-vulnerability safety measures in manual and autonomous cars are worthy of greater attention than currently received. Interesting as a means of promoting safety, critical attention should be given to matters of mediation and intentionality. While biomimetics and biosemiosis are based on natural systems, the mediation of people and things in autonomous and auto-vulnerable systems are not.

Hybrid Work

Automated for the People?

The modern world of work involves automation of all sorts of activities, such as the creation of job ads, job hunting, assessing applications, interviewing, selecting, monitoring, managing, disciplining, rewarding, training, and keeping a well-disposed workforce. This workforce may be in a fixed location, entirely mobile, at home, or a combination of these. The nature of work is also undergoing change through introduction of what this book designates automated empathy. As this chapter will explore, this is psychological, involving use of quantity to gauge qualities of temperament, behaviour, performance, accordance with corporate values, and individual and collective mental well-being. At least one risk is clear upfront: intensification of the psychological dimension of work as workers have not only their physical but also emotional performance gauged. Proponents of these systems argue that automated empathy may help manage, alleviate, and improve worker experiences of the workplace. Others argue that use of emotion-based analytics represent a boundary that should not be crossed (Moore 2018). Such concern is amplified by broader exponential increase of data processing about employees in the workplace in relation to location, behaviour, performance, suspected fraud, suspicious behaviour, health, social media usage, productivity, target setting, rewards, punishments, ranking, gamification, and in-work communications (Christl 2021).

Automating Empathy. Andrew McStay, Oxford University Press. © Oxford University Press 2024.
DOI: 10.1093/oso/9780197615546.003.0008

Beyond macrolevel concerns are worker experiences of automated empathy in the workplace. This chapter sees such testament as ethically important, but also intellectually useful to advance theorisation of what is taking place and what the significance of these experiences are. This chapter finds a bleak form of empathy gameplay at work, one where workers are required to feel-into the disposition of the workplace technologies to respond to organisational wants. Paradoxically, this denaturalises empathy and emotion expression at work, requiring workers to themselves simulate empathy and adopt hyperreal emoting behaviour. Yet the politics are not straightforward. While there is a growing body of work-focused academic literature concerned with impacts on marginalised groups (*unnatural mediation*) and the psychology of workers (*mental integrity*), and law that will likely ban biometric emotion profiling at work (the European Union's draft AI Act), datafication of workers *does* receive tentative interest from perhaps the most surprising of places: worker unions. Their interest is that transparency may serve unions (worker collectives) as well as human resources and organisational management teams. This is a hybrid ethics issue: one involving terms of connection and access. The final section of the chapter considers whether automated empathy may work for employees, serving to protect them from managerial misbehaviour. Admittedly, on beginning work for this chapter, this was not a proposition I expected to consider. However, as we have seen earlier in the book, hybrid questioning can lead to unexpected answers.

AUTOMATING INDUSTRIAL PSYCHOLOGY

Managerial use of data about emotions and affective states in the workplace is an attempt to make human industrial processes transparent, potentially leaving workers vulnerable to new modes of discipline (Mantello et al. 2021). The controversial nature of this is such that in 2021 the Council of Europe under Convention 108 stated that the use of technologies to recruit people based on their moods, personality, or level of engagement in the workplace poses risks of great concern (Council of Europe 2021: 5). Soon after, Recital 36 of the European Union's draft AI Act described that AI usage for recruitment and the workplace (including biometric data) is high risk because 'such systems may perpetuate historical patterns of discrimination, for example against women, certain age groups, persons with disabilities, or persons of certain racial or ethnic origins or sexual orientation' (European Commission 2023). However, AI used to detect emotions through biometrics receives the next level of treatment, as Recital 26c of the draft Act states that AI used in the workplace to 'detect the emotional state of individuals should be prohibited', a point that has two weaknesses. First is that it leaves scope for debate about use of emotion recognition at an aggregate level of the workforce, rather than individual workers. Second is that phrasing of detecting emotional 'state', rather than 'expressions', also leaves space for legal ambiguity. After all, managers may not care about an individual's emotional states, only how a worker outwardly performs. Perversely, a prime reason for the ban of emotion recognition (unreliability) could assist with defence of its usage. A potential claim is that a company only seeks to understand worker behaviour and communication, not interior states, and that they would not use technologies to gauge emotional states due to their unreliability. Amazon for example are explicit that their systems are 'not a determination of the person's internal emotional state and should not be used in such a way' (Amazon AWS 2023). It is a short hop to the defence that communication rather than interior states are being gauged.

Although this chapter focuses on modern work contexts, note should be made of the stages before employment. A growing body of literature assesses how automated empathy technologies may be applied to job posts to assess 'gender tone' with Textio's 'gender tone meter' claiming to measure if language in job descriptions risks alienating applicants due to gendered language (Bogen and Rieke 2018). After sentiment-sensitive job ads there is also automation of the recruitment funnel, especially the approach exemplified by Pymetrics that uses games to measure cognitive and emotional attributes, including matching facial expressions to emotions.[1] HireVue (a firm that allows recruiters to algorithmically screen candidates through online video interviews) is especially well-studied, in part because of their explicit claims to mitigate in-person human racism and stereotyping. Indeed, making use of accusations that the AI recruitment industry pedals 'snake oil' (Narayanan 2021), HireVue responded by alleging the same of its critics. It argues that their 'structured' interviewing, which means online video interviewing with analytics conducted on the content of the video, is fairer than 'unstructured' in-person interviewing, particularly for women with caring roles and neurodivergent applicants (Parker 2021). Nonetheless, concerns about fairness and racial discrimination have been raised, concentrated around what Benjamin (2019) phrases as 'engineered inequity'. Sánchez-Monedero et al.'s (2020) study of HireVue is granular and nuanced, concluding it to be weak on intersectional bias mitigation. It also finds that while 'bias mitigation' may be suitable for the United States, it is not in other regions such as Europe where the 'right to an explanation' is stipulated by legislation.

Show Empathy

Closer to this chapter's focus, having landed the job, what of the workplace itself? Sales and teleworking environments are a prime site of interest for the automated empathy industry, through sentiment, voice, and facial analytics. Leading 'customer relationship management' giant Salesforce, for example, use Amazon Connect[2] to analyse transcripts of recorded calls for sentiment and emotion dynamics in sales and customer interaction. Similarly, Zoom's *Zoom IQ for Sales*[3] promises automated analysis of transcripts speech to allow for retrospective analysis, identification of emotional nuances in meetings, gauging of attentiveness and key message performance, detection of use of filler words (un, ahh, like, oh) that negatively impact on sales, and customer reactivity. Perhaps less is well known is Uniphore,[4] a well-funded Indian start-up (having raised $620 million US dollars[5]), that analyse faces during online calls to establish hyperreal basic emotion scores (happiness, surprise, anger, disgust, fear, sadness). They also include secondary emotion scores, with these secondary emotions being subemotions caused by primary emotions (fear, for example, branches into nervousness and horror; see Shaver et al. 1987). Uniphore's visualisation of conversational dynamics also includes unique measures for sales in the form of satisfaction, activation, and engagement. Alongside visual cues they also use voice analytics to assess vocal tone, involving speaking style, politeness, empathy, and speaker sentiment. US-based Cogito is also illustrative of automated empathy in the sales and customer relationship management workplace. To encourage empathy, etiquette, confidence, and efficiency, it scores whether the worker is speaking too quickly, whether caller and worker are talking over each other, call energy levels, and the degree of continuous speech, with each of these providing for an 'Empathy Cue' (described by Cogito as an 'opportunity to make a connection with the caller'). As per hyperreal and unnatural mediation observations made throughout this book, attractive and easy to read real-time empathy dashboards 'empower supervisors to virtually walk the floor with real-time measures of customer experience, agent behavior, and the ability to proactively listen to live calls [. . .]', with floor supervisors being 'automatically alerted to calls in which a customer is having a poor experience'.[6]

Powerful because it is mundane and relatable, a model reaction to unnatural mediation in automated empathy at work is provided by Angela, a teleworker for a contact centre. Interviewed by *The Verge* (Dzieza 2020) she describes her experience of being marked down by voice analysis software for negative emotions, despite other metrics being excellent. The software Angela was exposed to was from Voci.[7] Angela explains that she found her poor scores perplexing because her human managers had previously praised her empathetic manner on the phone. She says:

It makes me wonder if it's privileging fake empathy, sounding really chipper and being like, 'Oh, I'm sorry you're dealing with that,' said Angela, who asked to use a pseudonym out of fear of retribution.

'Feeling like the only appropriate way to display emotion is the way that the computer says, it feels very limiting.'

This leads to absurdity through hyperreal emotion prescription in that authentic human empathy is not only lost between teleworker and customer, but that Angela is forced to empathise with the system. From the point of view of understanding the significance of automated empathy, empathising with the technology is a significant point. Reminiscent of philosopher Michel de Certeau's (1988) interest in 'ways of operating', how to manipulate mechanisms of discipline, and how one should modulate and simulate conformity to evade them, Angela had to work out *its* disposition, intuitively understanding its black-boxed logics, and feeling-into its desired behaviours to please it, to ensure good empathy scores, and avoidance of unpleasantness with shift managers. Automation of call centre supervision is found with many other firms (such as Beyond Verbal and Empath) whose systems pertain to train people to articulate a range of speech styles (such as energetic, empathetic, or knowledgeable). Another aspect of fake empathy and human engagement is that systems could be tricked into better metrics (and lower caller empathy) by saying words such as 'I'm sorry' a lot, which would boost metrics. Notably, emotion and empathy in this line of work matters more than ever. Because bots are now able to handle the more straightforward cases (e.g., a password loss), this leaves the harder and potentially more involved human cases for call centre people (such as domestic or other emergencies). This of course is the time that one would like to speak to a person who can respond sincerely within the parameters of their job role. Thus, unique to social questions about automated empathy is the degree to which its installation involves loss of authenticity and human empathy itself from everyday life, especially in tricky circumstances requiring sophistication and care, rather than canned responses. The perversion is that the genuine human part of empathy risks being lost in the process of automating analysis of empathy, as workers are driven to game cybernetic empathy metrics.

Feeling-Into Meetings

A goal for those who employ data analytics in the workplace is to generate hitherto unseen meaning from patterns. Insights and observations can then assist decision-making. Microsoft's 'Workplace Analytics', for example, promises human resources managers an understanding about how busy a person is, productivity, effectiveness, emailing habits, meetings habits, network size within an organisation, how diverse that network is, the level of influence within an organisation, and whether that person might 'drive change' (Microsoft 2021). In 2020, under pressure from data protection activists such as Wolfie Christl,[8] Microsoft were forced into a public statement stating that productivity and influence scoring was only based on aggregated information, despite evidence to the contrary in previous versions of Microsoft Workplace Analytics that showed individual level capability. Early note should be made of claims to score 'only' based on aggregated information. As with previous use-case chapters, that point that a person may not be identified is not a good enough ethical argument to merit usage of automated empathy at work.

Linking well with transient data interest in Chapter 7 (George et al. 2019), this is problematic because privacy and data protection is typically seen as being about personal data (or information about a person or in connection with that person, which can single them out). Certainly, there is every right to be cautious that information collected under the auspice of aggregated insights might be used in relation to an individual, but the face-value claim is also important: what of the rights of groups (such as workers) not to be subject to methodologically unsound technologies that function in relation to intimate dimension of human life? Although this may make it legal in some regional contexts it does not make it right. Unnatural mediation, hyperreal, and mental integrity arguments apply, regardless of whether a person can be singled out or not. More generally, the sense that one's subjectivity is being processed and that one has in a panoptic fashion to be always-on, regardless of whether one is directly seen or not, is problematic. Also important is that judgements made on the basis of unnatural and hyperreal data *will* affect individual workers as collective decisions are made about groups composed of individuals.

Microsoft Workplace set the scene for other interests of Microsoft, including a patent filed on 30 July 2020. Titled 'Meeting Insight Computing System' the patent is for a system to understand and improve the productivity of meetings within an organisation by means of evaluating comfort, scheduling, and participant behaviour (United States Patent and Trademark Office 2020). From the point of view of the automated empathy thesis being advanced in this book, the Microsoft patent shows both capability and interest in scoring emotion expressions, physiological states, behaviour, participant histories, and connections with others, but also sensing of environmental factors such as temperature, air quality, noise, seating comfort, and room size in relation to number of participants. Although this book has a keen interest in physiology and emotion, the thesis is broader, encompassing all computational inferencing that pertains to simulate the feeling-into of everyday life. This includes biometrics but, like human empathy, it also includes proxies.

The parameters of quality for Microsoft comprise 'any piece of information that reflects, or can be correlated with, the efficiency, productivity, comfort level, etc., of a previously-elapsed or currently-ongoing meeting'. These are highly granular, including air temperature, air composition, brightness, ambient noise level, room occupancy, average attention level, average level of fatigue, emotional sentiment expressed by meeting participants (e.g., based on body language, tone of voice, explicit statements), percentage of participants who contributed during the meeting, attendance, materials distributed to participants in advance of the meeting (assessed by quality of the materials, how far in advance the materials were distributed, and relevance), whether the meeting started or ended on time, average number of other meetings attended by meeting participants that day/week/other period, results of a postmeeting survey on productivity of the meeting, participant roles or duties, participant workplace locations, distance travelled, whether the meeting is recurring, and how many participants attended remotely. In addition to analysis of ambient circumstances is presence and behaviour of contributors.

Meeting quality monitoring devices also watch for other behaviours in meetings, such as texting, checking email (and browsing the Internet), also benchmarking data against a participant's schedule and other meetings they have attended that day. Although there is often a big difference between a patent and what comes to pass, they are evidence of strategic interests and ambition, providing orientation (unguarded by public relations) to outsiders wanting a sense of how life-shaping companies are planning.

A good example of the potential ubiquity of automated empathy, Microsoft's patent is both alarmingly granular in its proposed surveillance yet also set to fail. There is first the problem of the sensing unwantedly interfering with the sensed, but it is likely to fail to capture communicational (especially meta-communicational) and interpersonal dimensions of meetings. Even the most positive work meetings are informed by contexts of hierarchy, power, existing relationships, institutional histories and, perhaps most importantly, what is unspoken in meetings. One is reminded of AI philosopher Hubert Dreyfus' (1991) long-standing warning that attempts to break down the context of human action into symbolically meaningful constituent parts is doomed to failure due to impossible complexity and the complexity of chains of meaning that sit behind an utterance, or expression.

Dispersed and Gig Labour

The place, nature, and arrangement of labour is changing radically. In the United Kingdom, for example, the percentage of workers paid by platforms such as Uber and Deliveroo increased from 6 percent in 2016 to 15 percent in 2021 (Butler 2021), a trend that shows no sign of slowing. Despite avowed empowerment though choice and self-management through gig work, the experience for many is one of precarity, worsened by having algorithms as bosses. In context of hybrid ethics (i.e., interested in terms of interaction with systems) one obvious risk is coercion and lack of respect for digital rights of workers.[9] Uber for example employs

Real Time ID (RTID) system in the United Kingdom, a system based on Microsoft's FACE API. Combining of facial recognition and location checking technologies, drivers must periodically take real-time selfies to continue using the Uber app (Worker Info Exchange 2021). Amazon also requires its drivers to sign a 'biometric consent' form to grant permission to access drivers' location, movement, telematics (such as braking, acceleration, speed, and phone usage), and biometric data in the form of facial recognition and in-cabin cameras (Gurley 2021). This was to authenticate identity and detect drowsiness and distraction. If drivers refused to give 'consent' their employment was terminated.

Although I discussed emotion, drowsiness, and distraction detection in Chapter 7, the use of vehicle behaviour as a proxy to 'feel-into' driver behaviour is important to highlight too. For drivers this is routine, but its mundaneness is telling regarding the creeping ubiquity of proxy-based automated empathy. For example, a report for the United Kingdom's Trades Union Congress (that represents 48 UK trade unions) illustrates this, citing one UK delivery driver whose vehicle registered high revs and hard braking, leading the worker to be accused of negative and aggressive driving (TUC 2020). On the face of things, perhaps this is true: high revs and hard breaking in an urban area might well signal irresponsible driving. Yet, the context is that the driver was working in the many large hills of the United Kingdom's Lake District. What is missed as well as what is logged is critical. Again, both a problem of unnatural mediation and a demonstration of need for a context imperative, objectivity paradoxically becomes relative to what a system can perceive. Even when the driver explained that these driving techniques were unavoidable, the system still marked them down for poor driving. The dynamics of power are all too clear given difficulties in challenging unfair machine-made decisions (especially if an employer claims that there has been a human review), computer insensitivity to context, mistrust of workers, inhumanity (by definition) and machine absence of care for human lives, and lack of worker control over data about them.

While context is important, we should again be mindful that context may be a solution that creates greater problems and harms. In context of work one can easily see how this would legitimise wider profiling of drivers, their cabins, and vehicles, in effect meaning that appeals to context would have an extra coercive character. This exemplifies a criticism that runs throughout this book in that critiques of automated empathy based on inaccuracy risk more invasive, granular, and personalised profiling. For drivers at work, surveillance legitimacy on basis of context is amplified by in-cabin cameras and sensors falling in price, so rather than recruiting properly and trusting workers, one can foresee context as legitimising extra worker surveillance (e.g., 'Tina, fair enough, we got it wrong. Management say we have to install these cameras to make sure we don't get it wrong again. Good, right?').

COVID-19 AND BEYOND

COVID-19 gave impetus to companies and employers seeking tools to keep tabs on employees, realised by November 2021 when global demand for employee surveillance software rose by 54 percent since the pandemic was declared. Although postpandemic homeworking began to be rebalanced with office-based environment, teleworking analytics and processes introduced at the crisis point will be hard to roll back. *Hubstaff*, for instance, is software installed on a personal computer to track mouse movements, keyboard strokes, and record the webpages visited on that device. Another, *Time Doctor*, downloads videos of employees' screens while they work and can access computer webcams to take a picture of employees every 10 minutes. Similarly, *TSheets* is a phone app developed to keep tabs on a worker's whereabouts during work hours. *Isaak*, made by Status Today, monitors interactions between employees to identify who collaborates the most, combining this data with information from personnel files to identify individuals who are 'change-makers', a premise that has potentially insidious implications. With online workers being less visible than those physically present in an office, one can see scope for the understandable promotion-hungry to game and perform for 'change-making' algorithms, potentially without making positive change to the workplace. Other software is more familiarly Taylorist in its surveillance, measuring how quickly employees complete different tasks, suggesting ways to speed up. This involves machine learning of workflow including what triggers, such as an email or a phone call, lead to what tasks and how long those tasks take to complete. Each of these are based on watching from afar with view to controlling behaviour (surveillance), but new approaches are emerging.

With empathy being about trying to feel-into and work out the disposition of others, the COVID-19 period saw a scramble for new media technologies that would allow people to connect, see, and interact as humanly as possible, but mass usage inevitably exposed the limitations of these platforms. The first few years of COVID-19 revealed a new phenomenon: 'Zoom fatigue'. With days increasingly populated by Zoom and Microsoft Teams calls, workers were forced through the medium of online 'camera-on' calls into a high amount of close-up eye contact. The problem was and still is excessive amounts of close-up eye gaze, cognitive load, increased self-evaluation from staring at video of oneself, and constraints on physical mobility (Bailenson 2021). This was found to be an intense and draining experience because unlike physical meetings where people may shift gaze and attention around the room and to objects (such as a notepad), with cameras-on everyone on a video call is paying attention to the faces and cues of others. The experience is subtle but affective in that the frame coerces people to behave as they would with intimate others, due to proximity of screened faces and extended direct eye gazing. One is locked into virtual walls and the gaze of others, but it also involves real-time self-inspection as one also watches oneself (the effect potentially being an infinity mirror of gaze-based self-inspection). With each camera-on call participant occupying their own digital cells, this gives rise to the strange phenomenon of visual intimacy at a distance. With people being close-up but far away, this leads to compensatory behaviour in the form of exaggerated agreement nods, thumbs up, and big end-of-call hand waves.

Absence of context, funnelling of attention and cognitive effort through a camera-eye sized window, and seeing oneself for an extended period is deeply unusual. This is especially so as workers are forced to simultaneously watch themselves perform, inform, make decisions, persuade, empathise, and support. Similarly, those whose jobs involves presenting, pitching, or otherwise persuading will know the ambiguous feeling of having performed as best as one can, given the limitations of the digital cell, yet are unable gauge reception and the reactions of others. Arlie Hochschild (1983) famously introduced the principle of 'emotional labour' to refer to extra levels of interpersonal and psychological effort in the performative and emotional dimension of work, such as with flight attendants and their performing of care and emotion, and management of passenger feelings such as nervousness in air turbulence. The introduction of complex gaze dynamics in video calls may involve similar question of empathetic exertion. With co-workers being close-up but far away, performative and empathetic exertion took the form of close attention to limited cues, but also having to handle how to be authentic in a digital cell that demands performative and inauthentic expression to achieve a resemblance of authentic in-person connection. Indeed, each big handwave at the end of a call is mutual recognition of low empathetic bandwidth and need to increase signal strength of parts of the communication spectrum than can reach the other.

Early in the pandemic, the limitations of video calls were recognised. Seeking to capitalise on the mediated empathy deficit, emotional AI companies such as Affectiva blogged solutions on how to 'read the room' through additional facial coding analytics.[10] Indeed, Cato Solutions' app *Smirkee* (that uses Affectiva's SDK) proposed to read expressions as a person speaks, showing the named emotion on screen. Mercifully this did not catch on, but the question of low empathetic bandwidth is one that more prominent companies continue to be interested in. Microsoft for example seeks to address the problem of limited access to audience responsiveness during online presentations (Murali et al. 2021). Noting aforementioned stress placed on online presenters due to lack of ability to gauge behavioural and emotional reactions, Microsoft's AffectiveSpotlight system analyses the facial responses (including emotion expressions) and head gestures (such as nodding in agreement or shaking in disagreement) of an audience, spotlighting for the speaker audience members who are expressive. With application to work and education, the goal is to increase sentic (emotion) and empathetic bandwidth to enable better communication between speakers and listeners by means of enhancing feedback to help presenters read a virtual room. In context of work, the lure is clear: use existing cameras and facial coding to gauge the emotional tone of the video call. One can see the proposition, to make calls and online interaction easier and reduce empathetic exertion by (1) broadening the spectrum of online communication, and (2) the quality of what may be communicated through channels with the spectrum. Yet, whereas Chapter 6 on education concluded with tentative acceptance of nonrecorded display-only mediation of emotion in online avatars, this is not what Microsoft's AffectiveSpotlight proposes. It would involve profiling, labelling, recording, and potentially (if not likely) scoring of worker reactivity and engagement levels. Should the latter seem unlikely consider that getting promoted was more difficult for online workers earlier in their career due to problem of decreased visibility compared to the in-person office environment (Morgan 2021). The risk is that worker surveillance is legitimised through worker opportunity to career progression, whereas early career progression would be better served by clearer organisational thresholds to recognise progression.

Feeling-into Remote Employees: Mixed Reality

With the so-called 'metaverse' (Ball 2020) promising an embodied internet as a means of enhancing presence, hype levels are high in the early 2020s for immersive technologies. Chapter 9 will consider fartherreaching claims about embodiment and technology, but Microsoft's 2021 announcement of 'Mesh for Teams' builds on the COVID-19 teleworking impetus and renewed interest in mixed reality. From the point of view of interest in problems of inter-personal empathy experienced during online calls and automated empathy, 'Mesh for Teams' is important because it seeks to solve how to create a better sense of co-presence between people who are physically distant. This is not at all new, with virtuality being long recognised as 'where you are when you're talking on the telephone' (Rucker et al. 1993: 78). Yet it is also substantially different in that co-presence through 'Mesh for Teams' indicates a substantial shift towards greater involvement of bodies in mediated communication and online work. The will to embodied virtuality through networked technologies has a surprisingly long history. Half a century ago, information theorist Ted Nelson's (1974) *Computer Lib* stated that we 'must design the media, design the molecules of our new water, and I believe the details of this design matter very deeply. They will be with us for a very long time, perhaps as long as man [sic] has left' (2003 [1974]: 306). The 'molecules of our new water' is useful phrasing, particularly for the datafication of mixed reality. Microsoft (2021a) for example articulate mixed reality in terms of 'instinctual interactions' that 'liberate' people from screen-bound experiences. Operating a continuum of augmented reality to virtual reality (AR/VR), workers will be represented in synthetic space in relation to their real bodies and expressions.

There is precedent in VR for in-world representation of expressions, for example through installation of electrodes in the foam facemask of VR headsets. Here sensors gauge levels of pressure applied to the sensors through facial muscle usage and changing of facial expressions (Li et al. 2017). Mass-market, Meta's Quest Pro (an AR and VR headset) uses inward-facing cameras to track eyes and facial expressions, enabling in-world facial expressions (also, at the time of writing late 2022, Apple's launch of a headset with similar expression-rendering capability was slated for 2023). The Meta headset is notable too because Meta and Microsoft collaborated to enable Microsoft services on Meta's Quest Pro. However, although AR and VR are increasing in popularity, Mesh for Teams was launched without need for headsets to smooth transition to the platform (Roach 2021). While its goal is 'holoportation' and photorealistic immersion in a synthetic space, it will also allow for phone and desktop cameras to mirror physical facial expressions for avatars in synthetic environments. The rendering of bodily behaviour and facial expressions in synthetic terms has multiple consequences. These include training of automated empathy systems, where facial expressions and physiological behaviour will provide limitless examples of face shapes for Microsoft. Beyond expression is relationships with synthetic context enabled by digital molecules, which provide gigantic data sets because biometrics, expressions, communications, and interactions (with objects and avatars) within that synthetic space can be quantified and experimented with. There is precedent for this in work with VR in the 2010s and early 2020s because retail experience designers, architects, interior and industrial designers, and market researchers have long used VR as a means of testing for affective responses and nudge-based reactions to in-world features (McStay 2018).

This book is sensitive to claims that the metaverse is more hype than substance, that it represents a desired outcome for Meta/Facebook rather than an inevitability, and that rhetoric is repeating itself given the history of Second Life in the early 2000s. The scale of investment in mixed-reality signals that there is more to it this time though. Microsoft have three prongs of relevant experience: their mixed-reality 'HoloLens' is widely used in manufacturing and engineering environments, giving Microsoft practical experience; they have invested significantly in emotion and empathy research (luring field leaders such as Daniel McDuff from the field-defining start-up, Affectiva); and they have long-standing expertise in workplace analytics. Especially when factoring for datafication of co-presence, expression, behaviour, language, and expression, the future of hybrid work looks to be one that tries to remedy the problem of being close-up but far away through biometric means, enhanced profiling, and by removing the walls of digital cells. Such high levels of datafication of work signal the need to consider what level of representation of bodies and behaviour is socially and morally acceptable in labour-based arrangements. The hybrid nature of this book is such that it can see utility in expansion of the empathy bandwidth to enable greater co-presence, including mediated emotion expressions. Concern begins when and if attempts are made to label, judge, and/or record behaviour.

FEELING-INTO SOLIDARITY: UNIONS

Whether in a fixed location, entirely mobile, or online at home, the various forms of work-based mediated empathy raise questions about rights and representation. A key stakeholder in representing individuals and collectives to achieve positive working conditions is the trade union. This involves genuine questions of ethics, especially those that recognise lived conditions of workers. Yet, as those who specialise in question of 'data justice' and labour recognise, abstract debate about ethics are worlds away from the lived experiences of overt and subtle power in the workplace (Dencik 2021). The dubious applicability of ethics is amplified when stances are less well-meaning, when ethics are a public relations exercise to legitimise desired or existing courses of organisational action. Placing ethics in the domain of politics, Dencik questions the absence of worker voices, unions, and labour perspectives in AI developments that will continue to affect them. She observes that policy frameworks tend to focus 'on the nature of technology rather than the conditions of injustice in which technology is situated' (2021: 1). While Dencik's grounding of ethics and justice in the lived experience of those subjected to coercive power and negative aspects of technologies is a good one, care must be taken not to introduce polarisation of ethics and justice. After all, justice is a central pillar of ethical, legal, and political philosophy (Rawls 1971, 1993). Yet, Dencik's urging for a ground-up worker-focused approach seems right, just as other groups represented in this book (such as children) have the right to be heard and for technologies to be used in their interests (or not at all). Dencik notes, however, that the labour movement has been slow to engage with questions of data and digital technologies, instead having focused on the material effects of these changes in the workplace in relation to experiences of work. This is a hybrid observation, one that looks to the specifics of technologies and lived terms of interaction in given workplace contexts. The consequence of this slowness to engage has for Dencik been an inability to challenge the functioning of the systems. This is not helped by the complexity of AI decision-making and opacity around both access to the systems and what they do.

If the labour movement has been slow to engage with questions of data and digital technologies, special mention should be made of aggregated automated empathy at work (noted above in discussion of Microsoft Workplace). When data collected about a person is (almost) instantly aggregated and cannot be linked back to a person, this invites question about whether it is still strictly personal data (data that relates to a person, or that might affect that person in some way). At work, while automated empathy data may be used to judge individual performance (as in the case of Angela above, the call centre worker), decisions may be made about groups and the workforce at large (e.g., empathetic and emotion performance of a room of 100 call centre workers). This leads one to privacy interests based on the group's identity, where groups should be recognised as possessing interests in controlling their group identity (Mittelstadt 2017). Unions may play a vital role here in negotiating and enforcing group rights, especially given the relative weakness of individual workers in arguing against collective workforce decisions that will certainly affect them, but only as part of an aggregate. The collective mental integrity principle introduced in Chapter 2 has bearing here. Spirited by phenomenology, privacy, and cognitive freedom, it is a dignifying principle. With dignity typically understood as the premise that 'every human being possesses intrinsic worth' (High-Level Expert Group on Artificial Intelligence 2018), a collective view simply drops 'every' to ensure that generalised human being should be respected. Backward note may also be made to Chapter 4 and its mention of Ubuntu ethics (that privileges community) and Kyoto School collective interests discussed in Chapter 5. These also sees individuals in terms of communitarian interconnectedness and communal existence. There is then diverse international ethical testimony that ethics may not be only applied to group interests, but that they should start with them.

Automated for the People

What is increasingly clear is that unions will need to become literate with analytics deployed in traditional workplaces and beyond. They will be required to challenge workplace analytics and performance management systems that pertain to 'feel-into' individual and collective attributes of workplaces (people and proxy technologies). Articulated positively, transparency and monitoring of emotion in the workplace may grant insight into whether employees are undergoing emotional and physical difficulties; identification of when people need support; mental and physical stresses; illnesses and conditions; and anger flashpoints. Yet most who have worked for an organisation will recognise that relations between workers and management can be adversarial. In theory though and with due caution, notwithstanding problems of unnatural mediation, privacy at work, forced consent, and marginalisation of worker interests, data about work may help surface problematic workplace conditions. For example, high aggregated stress levels across the workforce would show that there is a generalised problem. Such data is of potential value to unions, following the argument yet to be fleshed out in Chapter 11 that an inverted approach to automated empathy has potential for social good and encouragement of human flourishing (involving self-reporting, user annotation, and bottom-up learning rather than hyperreal classifications). Applied early here, with care, it grants scope for workplace transparency, enhanced collective bargaining processes, and better working conditions. Such suggestions however are fraught without worker representation regarding the use of technology and data at work. A hybrid view of automated empathy at work does not reject the premise outright, but the protocols (the nature of connection with systems and their managers) are crucial. Key factors are worker representation in relation to introduction of new technologies, a workforce that is aware of what takes place, giving workers appropriate levels of control over their data, and both educated managers and union leaders who understand and can respond to context-specific concerns.

These hybrid suggestions are echoed by the United Kingdom's Trades Union Congress (TUC), the federation of trade unions in England and Wales mentioned above. In their report on data and AI technology at work, the TUC find that workers were not aware of the nature of monitoring and AI decision-making, yet they were concerned about such processes that may materially affect them. This points to problems of disenfranchisement, lack of voice, alienation, and absence of feelings of control. Connected, the TUC's polling finds that 75 percent of UK workers sampled 'agree that employers should be legally required to consult and agree with workers any form of new monitoring they are planning to introduce before they can enforce it'.

Although not an endorsement of monitoring at work (far from), the TUC's findings is one based on dignity at work, appropriate manager-worker relations, and decent hybrid relations; a sort that admits of new technologies in the workplace yet demands that all stakeholders are served by them. Notable stipulates by the TUC are appropriate boundaries between work and personal life, transparency of usage, data accuracy, need for access to legal redress, rights to data protection (such as privacy and nondiscrimination) and timely intervention by regulators (TUC 2021). In return the TUC encourages closer engagement on use of new technologies between employers, trade unions, and workers. They also recommend closer collaboration between workers, trade unions, and employers, as well as technologists, regulators, and government. Such collaboration has chance of raising overall understanding about these technologies, identifying how all stakeholders may benefit, and when (and how) external agencies should intervene in case of abuses. Indeed, The Century Foundation, an influential US-based think-tank, believes that novel technologies and improved human condition are commensurable. Discussing the datafication of work and wellness in what is an otherwise highly critical report, it states that the problem with this development is *not* technological, stating that many workers can and would appreciate assistance to improve performance and manage their time. The key problem arises 'from a lack of worker input, control, or ability to contest the implementation of these devices' (Adler-Bell and Miller 2018). In the case of automated empathy at work, if deemed of value by all stakeholders (in the union's case, likely to help demonstrate problematic workplace conditions), a volte face in automated empathy values is required. This means flipping key fundamental premises of the values baked-in to automated empathy at work (see Table 8.1).

Table 8.1 Value Flipping: Creating Better Automated Empathy at Work

Passive	Active
Closed	Modifiable
Prescriptive	Self-annotation
Universal	Context-sensitive
Deductive	Inductive
Managerial	Collectively held
Mistrust of workers	Trust in workers
Robotised people	Mutual respect and community

The hybrid stance of the TUC is a good one in that it: (1) seeks to make organisational analytics work for all stakeholders; and (2) develop systems based on worker input to highlight problematic treatment when working. The TUC is not the only major body making this type of suggestion: UNI Global Union for example, that represents workers in 150 countries and negotiates with companies possessing a global reach, seeks to negotiate terms of usage in the workplace, not whether analytics in work should exist at all. Examining the future world of work, the UNI Global Union states that 'transparency is important because it builds trust in, and understanding of, the system, by providing a simple way for the user to understand what the system is doing and why' (The Future of World Work 2017: 6).

Care with Transparency

Transparency for the UNI Global Union does not mean access and view of entire systems, because transparency of complexity may hinder understanding. As a value, transparency is desirable, but it is if little help of what is seen is incomprehensible. Rather, a transparent AI system is one in which 'it is possible to discover how, and why, the system made a decision, or in the case of a robot, acted the way it did' (ibid.). Further, UNI Global Union state that workers must be able to access the terms of decisions and outcomes, as well as the underlying algorithms, thereby making systems and firms challengeable. Moreover, echoing the general recommendation made about the inversion of automated empathy, to be of a more bottom-up inductive sort, UNI Global Union state that 'Workers must be consulted on AI systems' implementation, development and deployment' (ibid.). Yet, care with transparency is urged, especially regarding the opening of institutional analytics to workers and unions. The UNI Global Union recognise that transparency may communicate little if one does not have the expertise to ask the right questions and understand given systems. While the word transparency has a rallying and 'enlightened' character, it is only of value if the information is not only visible, but comprehensible, and capable of being acted upon.

There is also the question of which form of transparency is being deliberated over. What the TUC and the UNI Global Union is advocating for can be interpreted as *liberal transparency*. This is that classical liberal and enlightenment norm which opens machinations of power for public inspection (Bentham 1834). The overarching principle is one of control so that the balance of power lies with a citizenry (Bakir and McStay 2016). This is the preferred stance of hybrid ethics, recognising the role of visibility, and that the *negotiated* terms of visibility arrangements are paramount. Yet, the sociotechnical goal of workforce management tools is historically *forced transparency* that opens the lives of workers to inspection. Transparency here means maximal visibility of people, but without their full knowledge or meaningful consent. An inversion of automated empathy (to user input/annotation) may also be too much of an anathema to be realistic, especially given the belief that automated empathy is typically done through passive means, where a person (and here, worker) is unaware of the nature of profiling. Management may argue for this under efficacy rather than malign intent. With self-reporting shunned since the 1960s when emphasis in psychology shifted from psychoanalytical and motivational techniques to observational forms based on behaviour (Satel and Lilienfeld 2013), the concern would be interference with data behaviours, responses, or reactions that people cannot or do not want to reveal. Between liberal and forced transparency, management might argue for *radical transparency*, basically meaning that if we were all more open, we would be happier. Identified by Jeremy Bentham, 'Every gesture, every turn of limb or feature, in those whose motions have a visible impact on the general happiness, will be noticed and marked down' (Bentham 1834: 101 cited in McStay 2014). In this situation, the *whole* workplace would become more open through use of tools and automated empathy analytics, on the understanding that greater transparency of management (physiology, affects such as stress, and datafied emotion) would lead to net organisational general happiness. Unrealistic due to historic opacity of management, the question remains of what happens if one does not wish to participate in a radical transparency arrangement? Radical quickly becomes forced transparency.

Souveillant Empathy

So far, we have assumed that use of automated empathy at work would involve systems installed by management (likely developed by third parties) and, in the most positive cases, this would be done through meaningful negotiation with worker unions. What though of systems chosen and perhaps by developed by workers or their representatives? The United Kingdom's TUC, the federation of major UK unions, sees value in the use of data in bargaining for better workplaces. This includes access to those installed in workplaces (in hope of progress through liberal transparency), but the TUC also promotes worker self-tracking through systems not installed by workplaces. This provides a twist to more familiar accounts of self-tracking applications, typically involving well-being initiatives from Human Resources departments that are exercises in self-optimisation and 'bossware' due to their tracking capabilities (Lomborg 2022). Worker self-tracking through systems that are not installed by workplaces is taken here to be souveillant empathy, where workers monitor their own time, motion, labour and well-being. This involves principles and politics of sousveillance, the act of watching ourselves and others partaking in the same activity as us; and the act of watching the surveillant powerholders (Mann et al. 2003, Bakir 2013). The TUC recommends use of Prospect's life-sharing app, *WeClock*. This allows workers to quantify their workday, ranging from the best paid to the most precarious and least well paid. In addition to tracking location and physical steps, it allows users to self-track unpaid labour, log inability to disconnect from work, report bad workplace practice, and register physical, mental, or personal strain. Datafication of work here may both help workers individually and collectively to hold management to account. The collective characteristic is where systems differ from hyperindividualised quantified-self principles, as *WeClock* combines data sets provided by its users to understand the overall scale of over-work, unrealistic expectations, and toll on workers. To log experience along a continuum of overworked to well, *WeClock* asks 'How do you feel?' offering ten emoji options, beginning with 'Loudly Crying Face' through to 'Face with Tears of Joy'.[11] Arguably itself hyperreal emotion, prescriptivism, and a caricature of experience, a key difference is that labels are self-selected by *WeClock* users. It is one based on a trade-off of time (tired workers will have little appetite for having to input sentences), literacy (emoji are recognisable and quick to learn), but also simplicity in that distillation of complex experience into simple labels is expedient for better workplace bargaining purposes. The app also asks, 'Were you paid for all your time today?' offering a yes/no response in the form of a dollar bag emoji if yes, and an angry emoji if no. The app then allows a free text 'anything to add' journal entry option to reflect on the day. Inputs are logged as a 'Journal Entry' containing data, time, source (e.g., iPhone), wellness (emoji type and marks out of 10), and whether the user was fully paid. It remains to be seen whether *WeClock* will become popular but the social principle in the design is important. With sousveillance being the practice of recording activities from the perspective of the participant in the activity surveilled, *WeClock* exemplifies this by aggregating sentiment data (derived from highly felt experiences) to be used to bargain for better workplaces.

CONCLUSION

As a testament to the hybrid approach of this book that advocates understanding terms of connection with technologies, benefits, downsides, and how they affect the existing order of things, this chapter ends with a conclusion not expected at the outset. Although automated empathy may be readily interpreted through the dystopian prism of iron cages and dehumanising scientific workplace management, this chapter concludes with cautious interest in use of technologies to gauge experience. The risks are clear, perhaps most tellingly articulated here in relation to proxy data generated though gig work drivers, where engine revs were taken as signs of aggression. This example is significant because of its everydayness and that it involves judgement and management by algorithms and lack of sensitivity to context. In automated empathy in a fixed workplace, Angela was perversely required to tactically empathise with systems designed to feel-into her voice emotion and energy levels. In theory and practice, this is absurd. If these examples portend what it is to work with automated empathy, there is reason to be concerned about systems at patent and pre-launch stages. Microsoft were shown to be highly active in planning for the future of work, initiatives given extra impetus by experience with COVID-19. Although it is hard to see how the full details of their patents could come to pass (seeking to score on-site emotion expressions, physiological states, behaviour, worker histories, connections with others, and environmental factors of temperature, air quality, noise, seating comfort, and room size), it is the direction of travel that is significant: one that argues surveillance of spaces and psychophysiology is positive. The chapter progressed to consider rights of groups and collectives. This is the terrain of worker union interest, which sees a coming together of too-often abstract ethical discussion with injustices on the proverbial shopfloor. This expected outcome might be a rejection of automated empathy analytics, but the result is something more hybrid in orientation. It is one where (with enhanced law, regulation, literacy, access, oversight, and control) workers and their unions may benefit, along with employers and commercial interests. This is a highly hybrid ethical conclusion, one that does not suggest banning technologies per se, but radically altering human relationships with them so they may serve rather than exploit people.

Waveforms of Human Intention

Towards Everyday Neurophenomenology

Automated empathy refers to use of computational technologies employed to identify affective states, such as emotions and cognitive states, and thereafter make judgements and decisions about an individual or group of people. Although it overlaps with emotional AI, this chapter proves the point that automated empathy is not synonymous with it, as it enquires into technologies that profile and interface with the human brain. Certainly, there is hype around brain–computer interfaces but policy and industry bodies such as the OECD and its Recommendation of the Council on Responsible Innovation in Neurotechnology take it seriously, seeing health and therapeutic value, but also privacy, consumer, and civic manipulation harms (OECD 2019). A prime concern regarding neurotechnologies is mental integrity. As some readers will be already aware, especially given Meta's and Elon Musk's Neuralink work in this area, direct communication with others by thought alone is not an impossible proposition. Indeed, with Article 5 of the European Union's draft AI Act seeking to prohibit 'putting into service or use of an AI system that deploys subliminal techniques beyond a person's consciousness', concern about the potential for technologies that interact in a direct way with the brain is high.

This chapter does not attend to uses in health and dedicated medical contexts (Glannon 2016), nor disability and assistive uses (Aas and

Automating Empathy. Andrew McStay, Oxford University Press. © Oxford University Press 2024.
DOI: 10.1093/oso/9780197615546.003.0009

Wasserman 2016). It focuses on implications of neurotechnologies (here-after neurotech) used for non-clinical or professional care activities. In other words, the focus is on technologies and applications that have undergone biomedicalisation. This is where medical technologies and thought are applied to matters outside of the medical field, and where life and social processes are explained though technologically enhanced medical terms (Clarke et al. 2010). With addition of corporate, governmental, and other nonmedical strategic interests, brain-based means of human-technical connections raise social and ethical questions. From today's vantage point biomedicalised usage of neurotech seems fanciful, an attention-grabbing topic of dystopian allure yet not one needing serious attention. This chapter sees differently, motivated by belief that if there are commercial and other organisational opportunities in engaging directly with the brain, development of nonmedical applications has chance of becoming prevalent. Perhaps not in the 2020s, but eventually.

Involving classical questions of mental privacy and data protection, data privacy overlaps with neuroethical investigations. Given that these ethical questions involve nothing less than engineering of human beings and experience, they matter. Neuroethics is concerned with the ethical questions that are emerging in relation to understandings of the brain, neurotech, and potential impacts on society (Marcus 2002). The brain is especially sensitive, perhaps foremost because it directly involves selfhood, perception and thought. While an onion model of the self is imperfect, it illustrates a protected inner shell, reached by peeling away outer layers of surface subjectivity. Given that recent decades have seen an exponential increase in gauging and measuring increasingly intimate behaviour for commercial and governmental reasons, one is forgiven for being queasy about direct measures and interaction with the brain.

This chapter outlines the historical context of online neurotech, biomedicalised usages, and ethical questions. It pays particular attention to difficulties with liberty-based arguments and so-called predictive uses of neurotech, illustrating that ethical questions rarely have easy answers if considered properly. The chapter concludes by assessing neurotech through the prism of 'neurophenomenology'. This is a term developed by

Francisco Varela (1996) to account for the bifurcation of 'neuro' (involving gaugeable brain activity) and phenomenology (the study of experience). Varela did not have in mind companies and organisations that would be interested in the connection between experience and brain, but this is increasingly important. Potentially at stake is not only intimate profiling, greater insight into first-person perspective, but potentially material simulation and pre-emption of human experience and decision-making.

Ultimately, the chapter argues that while there is much hype regarding biomedicalised potentials of lab-based Functional Magnetic Resonance Imaging (FMRI) systems, if even some of these claims are possible, they raise profound questions about the nature of human relationships with technologies, each other, companies, and governments. This is because nonlaboratory uses shift human behavioural profiling concerns from inferring of experience and subjectivity, to direct observation of inner experience and the electrics of preempting decisions.

WHEN FEELING-INTO BRAINS WAS NEW

As ever, a historical perspective both helps ground understanding of the technologies and the hopes and social discourse they informed and were informed by. Neuro-technologies have roots in the late 1800s, when Richard Caton used a 'galvanometer,' that detects and measure electrical activity on the brains of rabbits and monkeys to test reactions to different stimuli (Finger 2001; McStay 2018). In the 1900s, Hans Berger invented the electroencephalogram (giving the device its name), recording the first human EEG signals in 1924 and published his detection work on people in 1929 (Millett 2001). This was philosophically as well as technically important, preempting the affective materialism of modern discussion of emotion, disposition and even free will. With Berger having a philosophical interest in Baruch Spinoza (1996 [1677]) and monism (a single-substance view of the world that seeks to bridge mind-body dualism), Berger used the electroencephalogram to argue that cerebral blood flow, cortical metabolism, and emotional feeling of pleasure and pain, are part of the same psycho-physical relationship. On hopes and visions, Berger saw telepathy as possible (which must have been quite a claim in the early 1900s), because he believed human thought was endowed with physical properties and these, in theory, could be transmitted from person to person.

This has roots in the principle of 'univocity' that has origins in the late thirteenth century with Duns Scotus, who maintained that there is a single unified notion of being that applies to all substances and outcomes (McStay 2013). Such monism is however more commonly associated with Baruch Spinoza, certainly in philosophy and neuroscience literatures. In the 1600s Spinoza famously posited that the mind and body are not autonomous, but instead work in parallel (as with neuroscientific accounts). Spinoza (1996 [1677]) also goes further saying that mind and body are indivisible and are somehow made of the same substance and that physical or mental 'worlds' are extensions of one and the same substance, and all are equally subject to causation. Notably, given later sections on the 'problem' of free will in the neuroscientific worldview, Spinoza also rejected free will and choice. Single-substance views are today encapsulated in Elon Musk's remark at a Neuralink promotional event: 'What I find remarkable is that the universe started out as quarks and electrons, we'll call it like hydrogen, and then after a long time, or what seems to us like a long time to us, the hydrogen became sentient, it gradually got a whole more complex, and then, you know, we're basically hydrogen evolved, and somewhere along the way, hydrogen started talking and thought it was conscious' (Neuralink 2020). Regarding medical rather than philosophical interest in single-substance approaches to human life, in the seventeenth century the neurologist Thomas Willis mapped the body arguing that anatomy is the basis of human experience, not immaterial forces (Dixon 2012). The turn to biology also opened the way for later understanding of emotions as affective and biological, rather than immaterial. Notably, Willis was also interested in 'man-machines', transmission, reception, and communication within feedback systems, providing the seeds for cybernetic understanding and thereafter emotional AI and biofeedback.

'Man'-Computer Symbiosis

By the 1940s cybernetics had established itself. Although the word is synonymous with technology and many blockbuster movies, it has applicability to both machines and living things. Ultimately it is about systems (created and found) that adjust to circumstances through the process of feedback. These systems will function through communication conceived in terms of information rather than meaning. As a philosophical intervention its impact was profound because it saw little point in the concerns of humanities, philosophy or even 'thought' in general. Indeed, the same applies to our interests in subjectivity, emotions, and that these vague experiences are now subject to the empirics of quantification and informatisation. Whereas emotion recognition technologies are popularly criticised for not being able to understand emotions like people do, this was never the right type of question for cybernetics. For cybernetics and the information theory it inspired, information had no dimension of its own and no innate connection with meaning (Shannon 1976 [1938], also discussed in McStay 2011). Meaning was never the point: *fungibility* was. Claude Shannon with his close colleague Warren Weaver clarify, saying 'two messages, one of which is heavily loaded with meaning and the other of which is pure nonsense, can be exactly equivalent' (1949: 5). Information should not be confused with meaning. This is underlined by Norbert Wiener (1948) who popularised terms such as 'feedback', 'input' and 'output' to describe the actions of machine, humans, animals, and the nature of communication.

As noted in Chapter 5, philosophers such as Martin Heidegger responded badly to this outlook, declaring in *Der Spiegel* in 1966 that cybernetics had taken the place of philosophy (Zimmerman 1990), fearing that matters of subjectivity and being had been lost to understanding of people as systems. Yet, Norbert Weiner also recognised the temptation within cybernetics to turn to materialism and cognitivism, in effect belittling subjectivity and downplaying the role of culture, society, history, human feelings, and emotions. Of course, Weiner was more optimistic about the role of cybernetics in society seeing need to anticipate potential consequences but also urging scientists and engineers to engage 'the imaginative forward glance' with the 'full strength of our imagination' to assess the implications of a given development (1960: 1358).

The suggestion of brain-computer symbiosis itself is not at all new either, prefigured by Joseph Licklider (1960). Funded by the US Defense Advanced Research Projects Agency (DARPA) Licklider posited 'man'-computer coupling that he speculated to go some distance beyond contemporary human-mechanical machine relations and that 'in not too many years, human brains and computing machines will be coupled together very tightly' (1960: 4). Cybernetically seeing people and computers as information handling systems, Licklider's hope was that the computer part of the man-machine symbiont would handle the 'clerical or mechanical' parts of any intellectual task (such as searching for information, calculating, pattern-matching, relevance-recognising, plotting, transforming, determining consequences of assumptions or hypotheses), so 'preparing the way for a decision or an insight' (1960: 6). With what today would be phrased as a 'human-in-the-loop' relationship, Licklider sought to augment work rather than automate it. Although the idea of symbiosis seems to assume two conjoined elements, Licklider's thinking has more in common with what today is referred to as cloud-computing by means of 'a "thinking center" that will incorporate the functions of present-day libraries together with anticipated advances in information storage and retrieval and the symbiotic functions' (ibid: 7).

Applications of speculative human-computer interfaces to war were unsurprisingly early to feature. Licklider described problems of latencies in computing for real-time war operations, and the lack of intuitive interaction when making critical war-oriented decisions. Interest in neural code and human-machine interfaces continued. Journalist Michael Gross (2018) explains that DARPA has long sought to merge human beings and machines to create mind-controlled weapons. A key development was in 1997 when DARPA created the Controlled Biological Systems program. In interview with Gross, zoologist Alan S. Rudolph explained to him that the aim was 'to increase, if you will, the baud rate, or the cross-communication, between living and nonliving systems'. This entailed

Licklider-like questions of how to connect brain signals associated with movement, to control something outside of a person's body. Research on this topic that is in the public domain today takes place under DARPA's Neural Engineering System Design programme, which 'In addition to creating novel hardware and algorithms' conducts research 'to understand how various forms of neural sensing and actuation might improve restorative therapeutic outcomes' (Arthur n.d.). Indeed, neurotech and human augmentation is a prime interest for defence branches of governments, who seek to modify human performance. For UK defence and security this means optimising, enhancing, reducing stresses and impact of extreme climates, and restoring people (Defence and Security Accelerator 2022).

Whereas Licklider's vision was language-based, control of external objects using EEG signals began in the 1970s when Jacques Vidal (1973) described the Brain Computer Interface (BCI). Here, brain signals themselves are the language for 'man-machine dialogue' and elevation of 'the computer to a genuine prosthetic extension of the brain' (1973: 158). Finding that 'EEG waves contain usable concomitances of conscious and unconscious experiences' and that 'the set of continuous electric signals observed does not for the most part consist of random noise as was often suggested' (Ibid.: 164), this for Vidal created possibility for dialogue. Vidal also recognised the role of other physiological inputs to computers, such as eye movements, muscle potentials, galvanic skin reflex, heart rate, acoustic, and somato-sensory arousal information (such as pain, warmth, and pressure).

Interestingly too, Vidal cites Yourdon (1972) defining 'on-line' computer systems as 'when they accept input directly from the point where the input is generated and return the output directly to the point of consumption' (1973: 169). Although viewed as computer-to-computer networking, Vidal's point is that if people can give and receive input, they too may be online through 'man-machine interaction.' Indeed, online in Vidal's experiments meant through a computer plugged into the UCLA node of the ARPA Network (the packet-switching network that pre-dated the Internet). Notably too, a use case defined in Vidal's seminal paper is *Space War* (a video game developed in 1962), which has implications both in term of modern popular culture (gaming) and military applications. Vidal also described how 'subjects are given an opportunity to fire [through EEG signals] "missiles" at opponents' space ships' and the registering of evoked EEG signals through the 'explosion' of a ship on the display screen. The paper also discusses situation-specificity, reflecting on the absence of a stable neural code that can be carried from one session to the next. Instead, Vidal provided for 'a learning period during which the computer establishes its reference data,' a process which allows for changes in context and unique dynamics bound to a period of interaction (be this gaming or military neuro-based telepresence).

Waveforms of Human Intention

Although cybernetics has application to all complex systems that function in relation to their environment through feedback, there is a clear link to neuroscience, not least because Warren McCulloch (a founder of the cybernetics movement) was himself a neurophysiologist. With neuroscience being comprised of neurophysiology, psychology, and the understanding of neural phenomena in relation to the exercise of psychological capacities, along with questions about the possibility of incorporating human nature into science (Bennett et al. 2007), neuroscience is itself a broad topic. While it might appear as an attack on questions of consciousness and lived social life, this is not so. As per Scotus and Spinoza above, the reverse is true, the symbolic and physical are seen as one and the same, because people are physical systems which act on representations. Lived social and cultural life seen this way is not only real, but necessary for people to coordinate and communicate. The 'neuro' part of this chapter, then, specifically focuses on the use of neuroscience to: (1) create new relationships between people and machines, especially in relation to machines that are tools in some way; but also (2) scope for systems and others to profile people.

Neurological determinism is when actions, decisions to act, and overt bodily movements are the product solely of neurological (and other physiological) processes and events (Waller 2019: 19). The discovery of electroencephalography (EEG) in relation to brain activity that precedes self-initiated movement goes back to 1965, when Kornhuber and Deecke found 'readiness potential' in the form of activity in the motor cortex and supplementary motor area of the brain before voluntary muscle movement (Kornhuber and Deecke 2016). This is a significant point to digest, observation that there is brain and biological activity before a person has a conscious experience of choosing to act.

Influentially, Libet et al. (1983) later found that, on average, the reported time of first awareness of decision to move occurred almost a third of a second after the start of detection of readiness potential in the brain, thus indicating that choice comes after a decision has been determined. Claim of material change in the brain before a decision is a profound point. The implication of what has come to be called the Libet paradigm is that 'if our conscious intentions never cause our actions (epiphenomenalism), then we lack the control required to act freely and to be morally responsible for our actions' (Waller 2019: 22). This, if correct, means that the conscious experience of deciding to act comes late in the causal chain leading to action.

Further, our life choices, may not be made by conscious acts, but unconscious acts, also indicated by detection of readiness potential. The significance of Libet and readiness potential is hotly debated, for example in relation to whether it is the pre-conscious intention to act (make one choice over another), a general cognitive bias or preparation for a voluntary task, which means that decision is not yet settled regarding which action will be taken. Yet, what is clear through multiple post-Libet studies is that some choices *can* be predicted, if a person is wired-up, granting around potentially one second of time between signals of detection of readiness potential, to conscious act (Maoz et al. 2019).

Extending this, studying the activity of the frontal and parietal cortex, a Libet-minded group coordinated by Soon et al. (2008) reflected upon their earlier work that also found that the outcome of a decision can be encoded in brain activity before it enters a person's awareness. This again has implications for self-governance debates (who and what is in control of a person). It is technically controversial, because the latency here was found to be only a few hundred milliseconds. This led critics to argue that the short delay between brain activity and conscious intention leaves ambiguity whether the prediction of intention really takes place (2008: 543). To factor for this Soon et al. inserted choice into their research, asking participants to choose between options (in the form of left and right buttons and letters on a screen), act (by pressing buttons) and remember (button and letter combinations). Due to being able to choose a button when participants felt the urge to do so, this granted opportunity to examine the period between changes in behaviour in regions of the brain, and the conscious act to click. In context of the controlled study involving buttons, screens and choice selection, Soon et al. found pre-conscious to conscious latency to be *seconds* long, in some samples by up to 10 seconds long. They also found that the period between conscious intention and clicking was on average 1 second.

It is highly likely that in the non-controlled setting of everyday life, with other incoming information competing for attention, that seconds would reduce back to milliseconds, but what does a world look like where human intention is not only pre-empted by our own brain, but objects, software and environments react to pre-conscious decisions? The answer has at least two elements: temporal and decisional parts. While a lag of seconds would involve a person forever catching-up with themselves, a latency of only milliseconds between the neuronal event and the arrival of the phenomenal decision would presumably feel like immediate interactions with technologies (such as games, cars, or military drones), due to the sense of immediacy being guaranteed by reactivity of a system to the brain's pre-conscious determination. (This assumes that all other parts of a technical system are functioning properly and without lag time of their own.)

The second part, decisional control, is also controversial due to its connections with autonomy and self-determination, distilling to the matter of whether actions are freely willed or not. The Libet paradigm suggests not, with multiple studies pointing to prediction rates of around 80% regarding how one will act prior to a what person says or does (Maoz et al. 2012; Salvaris and Haggard 2014; Waller 2019). Significantly, 80%

does not equate to determinism. This is important in that there is a correlational relationship, but not determinism. It should also be noted that these predictive studies are very short-term predictions, able to claim nothing for longer term decision-making and planning. Indeed, the Libet paradigm (predicting button click choices) draws its evidence from actions that lack consequence and purpose. This is key 'because it is deliberate, rather than arbitrary, decisions that are at the center of philosophical arguments about free will' (Maoz et al. (2019). This is backed-up with waveform studies, finding that 'the RP was altogether absent—or at least substantially diminished—for *deliberate decisions*' (Ibid.), although the authors do not claim that consciousness is more involved in deliberate decisions.

We have of course been here before, with Freudian determinism being founded on the premise of the unconscious mind that determines personality and behaviour. Questions of neurological decisional control also raises questions about the ability to act freely and the extent to which a person should be held responsible for actions (be this to praise or punish them), or their behaviour (good or bad) is involuntary. It also raises questions of what self-governance and 'freely' means, given that changes in personal, social and physical environmental circumstances may alter actions and decisions made. As Mele (2014a) argues, context matters, in that 'freely' means that there are options open to a person, given all circumstantial factors that contribute to a situation requiring a decision. Connected, there is also the 'compatabilist' argument that asserts that potential determinism and free will (and therefore moral agency) *are* compatible with factors that determine and orient a person. This point goes back to Schopenhauer (1960 [1841]) who observed that that a person may 'will' to go left or to go right and that willed decision is their own. What they cannot control for is independence of the decision and impact of external factors. Determinism itself is not arguably the key issue, but when (and if) determinism *bypasses* agential choice (Waller 2019).

On the philosophy of free will literature, neurological determinism can be seen in somewhat bleak terms in that free will does not exist because it is simply a set of neuronal events. This would allow for the happenstance experience of mental events, but these events themselves are not part of the material chain that effect change. Conversely, a neurological basis for will is a statement of the obvious: of course, the brain plays a role in decision-making, with readiness potential simply being part of human action. Indeed, it is often neuroscience that sets up the strawman dual-substance view that few humanist critics actually argue for (Mele 2014b). Perhaps most significant for concerns about determinism and remote prediction of decisions is that study of readiness potential and Libet are based on arbitrary and inconsequential decisions, rather than decisions that have consequences, external influences, and calculation of implications (Maoz et al. 2019).

Imaging Thoughts and Electric Dreams

Through use of fMRI scanners (that are very large, unwieldy, and expensive) there is increasing evidence of being able to remotely sense, infer and represent the classes of objects in thoughts and dreams. Building on work where scanners may be used to understand the binary options of what a person is thinking (such as black or white), Horikawa and Kamitani (2017a, 2017b) show evidence of being able to decode more sophisticated features of images, such as colour, structure, and feature relationships, for example allowing images of jets, birds or skyscrapers. In the case of imagination, Horikawa and Kamitani (2017a) measured brain activity while a subject was looking at an image. In other cases, brain activity was logged later, when subjects were asked to think of the image they were previously shown. With brain activity scanned, a computer reverse-engineered (or decoded) the information to create an image of the participant's thought. The video results[1] of the images reconstructed from brain data are certainly not at all clear, often appearing as impressionist blurs, but with effort some of the reconstructed images can be discerned (such as a doorknob). Letters of the alphabet were however clearly reconstructed.

In the case of neural representations of visual experience in dreams, dreams are not seen or directly observed, but again the objects within dreams may be predicted from brain activity during sleep. The premise is that brain activity patterns are linked with objects and scenes from subjects' dream reports, allowing for labels to be created in reference to neural patterns. Horikawa and Kamitani (2017b) measured fMRI and electroencephalography (EEG) signals while participants slept in a scanner, were woken when EEG signals and eye movements indicated they were dreaming, and subsequently asked to verbally report the content and objects within the dream. Similarly, Nishimoto et al. (2011) decoded and reconstructed people's dynamic visual experiences from Hollywood movie trailers by tracking blood flow through the visual cortex, although it is instructive to note that the procedure required volunteers to remain still inside the MRI scanner for hours at a time. In the long-term, however, the ability for neurotech to create visual portrayals of thought is profound. Yet, any suggestion that there exists a formal universal language of brain patterns and imagined objects is incorrect. Brain activity patterns are unique, like fingerprints (Finn et al. 2015), meaning that brain reading systems will need to be hyperlocal, likely requiring awareness and input from users of a system.

With a high-level implication being that neurotech image reconstruction is a textbook example of mediated and technical empathy, the practical applications are endless, especially with this data being online and communicable. Novel forms of art, online communication, media, gaming, surveillance, police evidence sourcing, and more, become conceivable. Crude and base, but true and real, it should not be missed that the pornography industry has a historical knack of utilising new technologies, be this in the form of home delivered video cassettes, video compression to fit online bandwidth, experimentation with virtual reality, meaning that direct interaction with the brain is has strong chance of being applied here first.

BIOMEDICALISING MIND-READING

While the studies cited here involve the use of expensive and cumbersome machines (such as fMRI scanners), recent years have seen interest, patents, initiatives, and investment in human-machine interfaces. Reinvigorating longstanding futurist and cyberpunk enthrallment to the 'Man-Machine' companies such as Neuralink and Meta (along with others such as Kernel, Synchron, Neurable, CereGate, and Mindmaze) are designing both invasive and noninvasive neurotechnologies for the commercial sector. How seriously should we take these initiatives? Are they to remain technologies only discussed in the outer reaches of techno-existential anxiety, surveillance studies, science fiction, and patent applications by 'founders' with limitless funds? Or, as this chapter does, should we see a modicum of inevitability in the use of brain behaviour profiling as a means of interacting with objects, environments, and each other? It seems to me that one may either relegate far-reaching claims for neurotech to founder fancy or take the situation more seriously, by seeing that the world's richest people have intentions for the brain itself, prethought, and imaging of imagination and experience. There are also factors of timeline and strategic intention in that, for the near future, yes, ubiquitous presence of such neurotech may be categorised under the heading of dystopian imaginary, but what of decades hence? Next is strategy, in that interest in such strata of human being is significant now. Although futures rarely arrive as planned this book takes these claims seriously.

Meta and Mixed Reality

In 2021 Meta claimed that augmented reality (AR) and virtual reality (VR) will 'become as universal and essential as smartphones and personal computers are today' and that they will involve 'optics and displays, computer vision, audio, graphics, brain-computer interface, haptic interaction, full body tracking, perception science, and true telepresence' (Tech at Meta 2021a). This vision of telepresence was a metaverse vision, a performative but costly one for Meta, corrected in 2023 when Meta promised investors that it would reduce spending on its Reality Labs. What is clear however is however one defines mixed reality, metaverses, or *the* Metaverse, the ambition for these is not new. Indeed, from the 1980s and certainly the 1990s, one can speak of a perma-emergence of VR and other immersive media that are genuinely powerful and different yet have not yet caught on despite multiple attempts at marketing them. While one must be wary of being swayed by future visions designed to rally investors and create industry networks and alliances (Liao 2018), and corporate prophesy of technological progress (Haupt 2021), this book takes the view that amongst the extraordinary levels of hype around mixed reality (and the metaverse) in the early 2020s, something important *is* taking place regarding the expansion of connections between people and digital environments. This involves a broadening of how people interact with content and digital objects, each other, with synthetic others generated by large language models, and more diverse personal data inputs into these interactions.

With Meta seeing (or prophesising) mixed reality as inextricable from the future of work, automated empathy in this context means sensing and measuring, or feeling-into, electrical impulses (such as through electromyography) in the body to gauge intention. In context of engagement with objects through VR and AR glasses, Meta explains their choice of a wrist-based wearable, where signals travel from the spinal cord to the hand, to control the functions of a device based on signal decoding at the wrist (Tech at Meta 2021b). Critically, the signals through the wrist can detect finger motion of just one millimetre. They also speculate (without mention of Libet's readiness potential) that 'ultimately it may even be possible to sense just the intention to move a finger.'

In what Andrejevic (2020) might refer to as a 'frameless' confine, while seeing privacy concerns the neuroethicist will also be interested in neuroenhancement and human augmentation (Lin and Allhoff 2008). With neural input technology moving towards everyday experience, there is clear scope to create, work, and track engagement with virtual objects. Automated empathy in this context means sensing and measuring (or feeling-into) electrical impulses through electromyography to gauge intention. The next part of this, overlapping with experience, is concern with personalisation and context sensitivity. Volume of privacy alarm in biometrically enabled mixed realities is increased when one considers that a principal challenge for Meta, monitoring online hatred and false information, would require close, real-time, and persistent scrutiny so harmful actors and content can be removed or neutralised as a quickly as possible. The risk is that the playing-off of risks (privacy versus on-line safety) against each other legitimises courses of action that best suit Meta, which are those that lock people into its services to surveille them in granular terms. The likely policy argument of 'How can we provide safety without listening' must be rejected on basis that it is an unaccept-able solution.

Meta also seeks to build 'adaptive interfaces' that function through inferences and judgements made about need, want and intention, in rela-tion to circumstances and surroundings. Surely with readiness potential in mind, the goal for Meta is that 'the right thing may one day happen without you having to do anything at all.' If the case for this as an ex-ample of automated industrial empathy were not clear, Meta promise, 'The glasses will see and hear the world from your perspective, just as you do, so they will have vastly more personal context than any previous interface has ever had' (Tech at Meta 2021b). One can debate technology, effective-ness, likelihood, and merits, but what should not be missed is intention— the strategic attempt for what will be argued as a neurophenomenology of human life, one involving both biology, experience, and mental integ-rity concerns. Given Meta's current business model, using technology to psychologically and sociologically profile individuals and groups to target advertising and sponsored communication, any revulsion would be per-fectly understandable.

Immediation: The Case of Neuralink

One of Elon Musk's companies, Neuralink, is building transhumanist technology that uses neurosurgical robots to implant groups of minuscule flexible electrode 'threads' into the human brain. This draws upon, but miniaturises, deep brain stimulation (DBS) technologies that are used to treat conditions including epilepsy and Parkinson's disease. Musk claims that Neuralink will solve a myriad of human health conditions. More central to this chapter is claim of eventual ability to enable superhuman cognition so people may coexist with advanced AI to achieve symbiosis and collectively stave-off an existential threat to humankind (Neuralink 2020). This is the territory of long-termism, or the moral obligations required to protect tomorrow's civilisations (Bostrom and Ord 2006; Torres 2022). While precautionary obligations are a good thing, they become more problematic when considered in reference to civilisations thousands and potentially millions of years hence, and likely not human as we know it today. It is a speculative form of utility theory, one where an imagined end (in this case general purpose AI) justifies painful means (need for brain chips so people can coexist with intelligences that would otherwise be beyond them).

Coin-sized, the Link itself replaces a piece of extracted skull with the device's 1,024 electrodes connecting to the brain itself. These threads, currently around a fifth of a human hair thick (5micromns), are inserted into areas of the brain, approximately 3 mm or 4 mm under the critical surface. Links read the brain by sensing electrical activity of firing neurons and write to the brain by a controlled electrical charge from the device's electrodes that are currently attached to the cortical surface of the brain (embedded 3 or 4mm under the surface), although Neuralink plan to embed Links deeper into the brain to enable interfaces with more brain regions. With ongoing testing in pigs (that bizarrely are claimed to be consenting volunteers[2]), in June 2023 Neuralink received regulatory approval from the US Food and Drug Administration to conduct the first clinical trial with people. Link threads and probes therein detect, amplify, and record the electrical signals in the brain. Having recorded them this allows for the transmission of the information by means of Bluetooth protocol outside the body. Like wearables, Neuralink point to being able to control a Neuralink app on iOS devices, keyboard, and mouse to 'directly with the activity of your brain, just by thinking about it' (Neuralink 2022). In reverse, Neuralink promises that that their system will write information to the brain, such as to restore the sense of touch or the act of vision, even providing new signals to the brain from sensors, giving 'superhuman' ability to see parts of the frequency range that people cannot usually see.

Telepathy is also promised, to save on effort of having to compress thoughts into words. Musk remarks that energy is wasted converting and compressing thoughts into words, that 'have a very low data rate' (Neuralink 2020: 1.05). Proposing to bypass encoding and decoding through language with telepathy, Musk says that speech is 'so very slow' and 'we can actually send the true thoughts, we can have far better communication, because we can convey the actual concepts, the actual thoughts uncompressed to somebody else'. Similarly, the act of *rendering* images or complex engineering models in the mind could be bypassed, by stimulating the correct arrangement of neurons in another person's brain. The goal then is a high-bandwidth brain-machine interface (BMI) system that effectively nullifies latency between the brain and technologies because the brain would respond quicker than the conscious mind (and its weak ability to use tools with buttons and icons).

Tellingly, Neuralink's Summer 2020 Progress Update refers to their service as a platform, meaning that it is a foundation upon which services and applications can be built (Neuralink 2020). This potentially creates opportunity for third party APIs, tracking device IDs and profit models based on network effects. Here the value of a platform is judged in terms of the number of people and organisations that use that platform. The more users there are, the more valuable the platform becomes to the owner, and its users, because of increased access to the network of users. Next is profit in that it is unlikely that neuro-tech platforms would restrict profit income solely to the sale of the Link itself. Rather, potential exists in allowing third-party companies to build applications for that platform, or to allow limited access to a person's device in the form of an API. If such a platform was ever realised, undoubtedly what could be read (and especially what could be written to the brain) would be intended to be limited and subject to careful controls. Yet, stakes would be extraordinarily high, and one only must think of Cambridge Analytica's exploitation of Facebook's now defunct data portal ('Friends API') that enabled access to not only users' data but that of all their friends with view to voter profiling and targeting (Bakir and McStay 2022). Indeed, if one can agree that it is a useful heuristic to consider that human exploitation will occur wherever there is scope for influence and profit, there will be many well-funded and organised actors seeking to exploit Neuralink's platform.

BECOMING EVERYDAY

It remains to be seen whether healthy people will allow an automated saw to carve out a circle of bone from their skull so a 'Link' can be installed (Elon, you first, please). Yet, less cranially invasive use of brain waves as an input and safety control system seems more feasible. Indeed, the *principle* of prostheses is an old one as tools have long enhanced and geographically extended human capacities. The concern is the terms of neuro-based relationships between people, technology and organisations, and how these are governed.

As discussed in Chapter 7 once unthinkable applications of computer vision to gauge fatigue and emotions are being installed in cars. This overlaps with this chapter's interests as the automotive firm Jaguar Land Rover (2015) has for some years researched in-cabin sensing of brainwaves that might indicate whether a driver is beginning to daydream, or feeling sleepy, whilst driving. Derived from NASA and the US bobsleigh team to enhance concentration and focus, Jaguar Land Rover's 'Mind Sense' research detects brainwaves through the hands via sensors embedded in the steering wheel, using actuators and software to amplify brainwave signals and filter out disturbances that might interfere with signals from the brain. In the case of autonomous driving, passive analysis would be required to gauge whether a driver is alert and able to take back control, requiring health, fatigue, sleepiness, and stress analysis, to ensure the driver is ready to take over. With data processing potentially restricted to the 'edge', alertness thresholds could conceivably see brainwave analysis become routine (a car would pull over if driving is otherwise unsafe). Indeed, given increasing prevalence of cameras in cars described in Chapter 7, falling prices of computer vision systems, and scope for Light Detection and Ranging (LiDAR) systems to detect blood flow and oxygenation in brains (Velasco 2021), one can readily see scope for both passive in-cabin surveillance and novel in-car system input. This would facilitate neuroergonomic interfaces, systems that adapt based on the characteristics of the neural outputs or the brain physiology.

All sorts of issues are raised, not least *authentication*: brain-based technology would require the strongest security protections, meaning that for purposes of mental security and privacy there would be a need to identify and authenticate one device to another. This would require mandatory device IDs, therefore opening brain-computer systems to public mass surveillance. This would be of the overt sort in regions such as China and Saudi Arabia, but also the covert sort such as that conducted under the auspice of 'Five-Eyes' intelligence-sharing arrangements for the purposes of national security. At a citizen and social level, it remains to be studied what the conventions would be regarding the sharing of personal details and contact means. Despite its ubiquity, a mobile telephone number is important personal data because it provides means of direct contact wherever a person happens to be. A neurotech ID of the brain-computer interface sort would likewise provide for highly intimate and experientially unmediated connections with others, inviting close attention to the conditions by which a person would make themselves available to another.

Slightly more subtle is *tool-being*, in that relationships with networked things are potentially altered. If the brain itself is effectively online, the relationship between a person and networked object changes as a person can interact directly with it. If ,for example, the Boston Dynamic 'Dogs' were not alarming enough, neurotech provides means of controlling these remotely, be this on nearby streets (policing) or the other side of the world (military). Less alarming, in theory, for citizens every object interacted with that uses electricity could be interacted with in more *immediate* terms (kettles, games systems, drones, and so on), potentially in a *touchless* capacity. The significance of this is perhaps seen in Yuste et al.'s (2017) citation of unpublished Google research that calculates that the average user touches their phone nearly one million times annually. Yuste et al. point to a qualitatively different media experience, where a direct link of people's brains to machine intelligence would bypass normal sensorimotor functions of brains and bodies used to [slowly] manipulate technologies.

Closely following novel means of engaging tools, is that *personalisation* of services, media content, and environments, would change dramatically, especially if neurotech were somewhere in the world to legally adopt platform logics as a business model that would provide for low-cost headwear, or even upfront invasive installation of the Neuralink sort. Akin to Amazon selling their Echo devices at cost price or below, this would be recouped by first party usage of data to personalise marketing (as with Meta's business model) or a third-party API ecology where limited access by platforms would be granted by a platform owner.

Cognitive Liberty

Many concerns about neurotech can be clustered under the right to freedom of thought, which in the neuroethics literature is phrased as cognitive liberty. Sententia (2004) argues cognitive liberty as the rights of individuals to mental independence, individual integrity, self-determination, autonomy, absence of coercion, and freedom from unwanted monitoring and interference. A glance from neuroethics to online media data ethics signals the role that third parties (in the form of browser trackers and APIs) might play, leading to similar concerns about mental data, third party access, and the potential degradation of personhood and personal autonomy (Ienca and Malgieri 2021).

Cognitive liberty and autonomy principles have a self-evident quality, but it is useful to stress-test these with specific situations that challenge all-encompassing normative outlooks. Lavazza (2018) for example cites a case where a person was treated for depression with closed loop deep brain stimulation (where electrodes are implanted into certain brain areas, generating electrical impulses that control for abnormal brain activity). Said person went to the funeral of a close friend but found that she was not sad for the loss. Simultaneously, the device 'read' the aggravation of depression and intervened to counteract it, in principle against the immediate 'will' of the funeral attendee. Certainly, the device user will have signed-up to this treatment, but the case serves to illustrate temporal complexities of will and agency that must be considered. Although this chapter is steering clear of medical neuroethics (by focusing on potentially ubiquitous applications), the deeply problematic nature of volition, temporality, first-person phenomenology, and consent in online contexts, is exponentially more complicated in context of both implanted and non-invasive commercial brain-computer interfaces (potentially for 'well-being' purposes).

In addition to questions of will are those of agency and identity in relation to neurotech, especially for brain-computer interfaces where systems might write to the brain itself. Given that liberty-based arguments draw power from Kant's argument about reason and a persisting and inquiring sense of self, what of proposals (such as Elon Musk's Neuralink) that seek to augment human processing with AI assisted processing. Where does identity lay? Perhaps instinctively one is inclined towards Aristotle's (1998 [350 BC]) *Metaphysics* that supposes subjectivity as an enduring primary substance of sorts, or a fundamental stratum which persists through change. This though has long been problematised due to deconstructive arguments about the role of language and culture in identity construction (Critchley 1999) and that memory has long been outsourced to books and digital storage. The stress-test here is about right to (and constitution of) subjectivity in context of neurotech. If the self *is* distributed and key areas of not only memory, but cognition, are occurring on third party propriety systems, this has scope to raise bizarre and perverse questions about who owns the stuff of selfhood (i.e., the establishing of insights and new information through creation of new links between neurons, be these augmented in the brain by neurotech, or on remote machines).

Another stress-test for mental integrity rights (i.e., cognitive liberty plus protection of first-person perspective) concerns applications that detect intergroup bias, including races, nations, ethnicities, political or religious beliefs. Would screening along these lines be acceptable in occupations such as policing that require just treatment all members of a community? This is *not* to advocate for such usage but to complicate strong normative stances. In their extensive review of fMRI studies of a range of human biases, Molenberghs and Louis (2018) find complex responses to race in the form of in-group and out-group bias (where one favours one's 'in-group'). They cite Cunningham et al. (2004) who found that when Caucasian participants watched black and white faces that were flashed very briefly (i.e., 30 milliseconds) they showed increased activation in the amygdala (the part of the brain important for acquisition and expression of a range of learned emotional responses) in response to the black faces. Yet, this effect disappeared when showed for longer, leading Cunningham et al. to conclude that participants were self-regulating the automatic implicit amygdala bias because participants did not want to be perceived as biased. Molenberghs and Louis (2018) go on to review similar studies, finding that fMRI studies of face perception show that people can regulate the increased amygdala response for outgroup faces. While Molenberghs and Louis (2018) reviewed amygdala responses in relation to favouring and in/out group bias, Chekroud et al. (2014) looked at racism through the prism of perceived threat. To add extra ethical complexity, they cite work that finds African American subjects showed greater amygdala activity to African American faces than lighter European American faces, leading Chekroud et al. to suggest that African Americans have negative implicit attitudes toward other African Americans. *If* amygdala responses are reliably found correlate with racial bias, this presents an applied ethical challenge to the sovereignty of liberty, through the right not to be discriminated against and to equal opportunity. Given that social expectations will lead individuals to publicly endorse egalitarian values that they do not privately endorse, especially if they want the job, one can see how grounds could occur to argue for legitimisation of brain scanning for evidence of social prejudice in work such as policing, education and perhaps beyond. As ever, the most important ethical questions do not lend well to binary good/bad answers.

Potential for neurophenomenological surveillance has parallel in ethical discussion of the polygraph lie detection, which in the US for example is used for security screenings, in criminal investigations, and as part of the conditions of release for sex offenders (Greely 2009). One could also see such scans as a standard background check where a person verifies that their brain will not favour or fear social groups or skin colours. Like polygraph lie detection, close attention must be given to efficacy. As Greely (2009) points out in assessment of fMRI-based lie detection, studies on deception have proved difficult to replicate, current work occurs in highly artificial settings, it is unclear which regions of the brain are associated with deception, it is deeply questionable whether deceptiveness signals can be averaged, and countermeasures may employed be to deceive the systems themselves. Moreover, although brain-inferencing is being developed by the world's wealthiest people (in 2022 Elon Musk was the richest person in the world), work so far on amygdala responses to racial bias is done through large and unwieldy fMRI scanners. We are then a long way from small systems capable of use for routine disposition scanning, although it is useful to preempt it and unpack the arguments that would likely be made in their favour.

Beyond concern about scope creep, efficacy and whether such systems can do what is claimed, there is the moral question of who gets to make the judgement about the social role of such technologies. Is it a middle-aged, European, and white academic with a background in privacy scholarship, such as the author of this book, or someone more at risk from social inequity? With the hybrid dimension of this book being interested in terms of interaction with technologies and those deploying them, and who is at risk, this is reminiscent of Nagel's (1986) discussion of moral objectivity and difficulty in establishing a moral 'view from nowhere'. In relation to ethics this tries to find a way through competing pressures, recognising 'agent-relative' concerns, but also trying to maintain ethical objectivity. Nagel sees that this cannot be straightforward, but usefully recognises that there 'can be no ethics without politics' (1986: 186). The upshot is that any ethical discussion must recognise plurality of perspectives within a social group, admit that individual perspectives cannot be transcended, yet paradoxically must also aim to be impersonal due to overall need for social (and legal) principles.

Neurobiology, Crime, and Violence

Another situation requiring ethical scrutiny is court evidence, which raises questions about mental privacy, permissible accuracy thresholds, and whether individuals should be held responsible for a crime on account of data about the brain? Some modern neuroscientists argue that criminal behaviour has a neurobiological basis, with brain scans having already been used in courts as evidence (Glenn and Raine 2014). This immediately raises accuracy concerns noted above, but also scepticism caused by the history of phrenology, physiognomy, and profiling of the body in policing (Thompson 2021). Proposition for more generalised use of invasive (such as Neuralink's) and noninvasive neurotech (such as Meta's sensors that read but are not inside the brain) will raise questions about admissibility of this data in courtrooms.

The legal context for interest in this data would be that in most cases a criminal offence requires both a criminal act (*actus reus*) and a criminal intention (*mens rea*), or the mental element of a crime. This includes factors such as conscious planning or intent, which in context of neuro-determinism are problematic notions because volition is a difficult premise if the brain is materially affected in some way. Straiton and Lake (2021) give the example of a schoolteacher who was convicted of child molestation and jailed. The night before entering prison the then ex-schoolteacher complained of severe headaches and was taken to the emergency room. An MRI scan revealed a tumour that was judged by the case neurologist to have caused the sexually charged behaviour, paedophilia, and inability to act on knowledge of what was right or wrong, likely due to interruptions of connections between the orbitofrontal lobe and the amygdala—the region of the brain responsible for emotion and decision making. Consequently, this was argued to result in diminished impulse control, affective impairment, and lack of feeling of right and wrong. Was the schoolteacher who had no known prior disposition to paedophilia to be blamed? The case calls into question fundamentals of agency, free will and an individual's accountability to society, given happenings to their brain. This is a rare case only uncovered because of coincidence with the

MRI scan, but more generalised usage of data from the brain in consumer-level brain-computer interfaces would surely lead to greater use of neuro-testimony. This not only raises questions about admissibility of this data but also about the nature of its significance. A situation is foreseeable then neuro-data where data from consumer-level devices is used to prosecute or defend on basis of responsibility and volition. From the teacher's point of view, it is highly likely they would want data from consumer-level devices to be admissible in court. Yet, the validity of brain data is called into question by observation that although there are correlations between anatomy and crime, it does not follow that a crime will be committed. Straiton and Lake (2021) give the example of war veterans who have taken blows to the head, who in turn show higher proxy levels for aggression than veterans who have not sustained injury. One can see grounds for neuro-established diminished responsibility, but this is problematised by findings of these anatomies in people who do not act on impulses to commit crime. Indeed, Glenn and Raine (2014) argue that it is problematic to argue diminished responsibility because not only are causal links are hard to prove, but all behaviour has a cause of some sort. This points to a view where biology can explain but not excuse, signalling that neuro-data may be used as testimony but not proof. More broadly, if a reality occurs where neuro-data becomes more generalised this requires closer consideration about the nature of responsibility, the role of environment other dimensions that influence crime, biological thresholds of proof and false positives, but also the relationship between punishment, public protection, and rehabilitation.

A Brief Question of Consent

Neuroethics is interested in guarding against unwanted access of 'the inner sphere of persons, from accessing their thoughts, modulating their emotions or manipulating their personal preferences' (Bublitz and Merkel 2014: 61). Although a specialist domain of ethics, much neuroethics is highly liberal in outlook. Bublitz and Merkel (2014) for example rely heavily on informed consent to protect the individual. This makes sense if neuroethics are being applied to health but less so when not. As has been repeatedly shown in media and technology ethics literatures, the practice of meaningful consent is problematic (Obar and Oeldorf-Hirsch 2020). Seemingly small factors, such as opaque wording, language choice, and button and icon design, all mitigate the principle that free and informed consent is straightforward in practice.

Much of this chapter has been speculative, trying to preempt life-sensitive situations that have chance of occurring. One might think that if such scenarios became seriously feasible then of course citizens would have stronger protections, given deep social sensitivities around the brain. It is useful in this regard to turn from what might happen to what exists now. While Neuralink and Meta's neurotech may not yet be present on the market, Muse is a leading wellbeing neurotech start-up. Their personal meditation assistant provides 'real-time feedback on brain activity, body movement, breathing patterns and heart rate during meditation'. Medically non-invasive and with far fewer sensors than promised by Neuralink or Meta, the Muse online privacy policy and end user license agreement is telling.[3] Reviewed on 04/05/21, the end user license agreement is 31,642 words long and, like many other examples of terms and conditions, it is written in opaque legal language. Putting the first few paragraphs through an online readability checker, it resulted in 'very hard to read' or 24.5 under the Flesch Reading Ease scoring system.[4] Such documents are designed to achieve legal corporate compliance than to obtain sincere permission from Muse users to process their data in given ways.

The Muse example raises the question of when should we become serious about protecting data about the brain? Better phrased, what is the nature of the threshold where special provisions might apply? Clearly there is large difference between seven external EEG sensors as with the Muse meditation assistant, versus that involving invasive surgery and removal of skull sections. In-between cases are on the horizon though, such as Meta's noninvasive thought-to-text services and their broader interest in hands-free devices that function through brain signals? Given that there will be a continuum of consumer-level neurotechnologies, how should they be judged and regulated? There are a variety of options, such as by sensors and sensitivity. One might also look to strict limitation of purpose regarding what is done with any processed data. Some are already looking to design-based solutions, such as Yuste et al. (2017) who argues that an ethical approach to these neurotech would be based on opt-in consent and federated machine learning. This flips the usual approach of sending data from a device for processing by algorithms elsewhere by downloading the algorithm, processing data on a user's device, and then sending back information about the algorithm (its weightings) that has been trained locally. One can see liberal appeal in decentralising approaches but there are more fundamental questions to be had about collective social safety, especially regarding what how non-personal aggregated brain may be used. One might instead conclude that an outright ban of consumer-level neurotech is the best way forward, but this too is problematic. It would involve banning of existing neurofeedback systems such as Muse that, despite its dubious approach to consent, is not especially socially challenging. Indeed, an overall ban would close services of potential home-therapeutic value to people. Views will differ on this but as the neurotech market grows there will be need for more focused governance-based investigation regarding not just scope for influence and untoward usage of data about the brain, but broader conversation about whether this should be processed at all beyond the strictly regulated remit of health.

A key reason for this conversation is that neurotech and data processing therein has potential to slip from view. If neurotech expands beyond the well-being sector to facilitate interaction with objects, content, and other people, it is likely that what initially seems outlandish and bizarre would quickly slip into the unnoticed flows and interactions of everyday life. Perhaps it is an over-reach to suggest that there is eventual chance that each mediated interaction will eventually involve traces of brain data about a person. Yet, the scale of financial interest in the consumer-level neurotech sector should not be ignored. Again, Elon Musk is the richest person on the planet as of 2022 and Meta has extraordinary planetary reach, given that as of the fourth quarter of 2021 Facebook had 2.91 billion monthly active users.[5] Neither are small start-ups who receive disproportionate amounts of media attention and criticism. Of course, the vast of majority of Meta users will not be queuing at product launch for a brain-computer interface. Given however that the goal of Meta's brain-to-text system is to naturalise communication and interaction between people this may eventually be an attractive proposition, if marketed well. Such intimate profiling would be neurophenomenological in character, involving the overlap between gaugeable brain activity and human experience.

WELCOME BACK FIRST-PERSON PERSPECTIVE

With the world's largest media and technology corporations seeing value in human-computer interfaces, this is a domain that has scope to develop under auspice of human-object-content-person immediacy. The reader is asked to temporarily resist go-to understandings of human excavation and mining of subjectivity (Heidegger, Autonomous Marxism, Surveillance Capitalism, etc.), post/transhumanist nightmares, cortex colonisation, technics, control, and biopower. They are asked to first consider that what is signalled is a profound engagement with subjectivity and inside-out first-person perspective. We might recollect here Meta's interest to 'see and hear the world from your perspective, just as you do' and their products 'will have vastly more personal context than any previous interface has ever had' (Tech at Meta 2021b). This inside-out strategy has a history, for example in Snap Inc's glasses and Google Glass. Ill-fated, interest in first-person perspectives of others (the inside-out gaze) should be noted, as should the additional neurotech aspect to these developments. Such strategy has a neurophenomenological character, with neurophenomenology being a diagnosis and theoretical attempt to redress the gap between first-hand experiences and neurophysiology. In short, neurophenomenology balances the biological and experiential sides of life (Varela 1996: 330). Might neurophenomenology represent a theoretical limit of surveillance, due to its interest in a person's interior modes of representation and brain activity, but also sense of presence, nowness, me-hood, and immediate horizon of experience?

The extent to which use of neurotech to profile, interpret, stimulate, and predict human experience is an ongoing topic for debate, but the direction of travel is to not only 'feel-into', but to effectively 'look out from'. This is to know in detail the material nature at the core of a person, to interact with their experience, and potentially decode imagined images and what a person has seen (Horikawa and Kamitani 2017a, 2017b). This puts us at the fringes at what is possible with large fMRI scanners never-mind smaller devices, but the strategy is an important one. It differs from the cybernetic behavioural logic underpinning much existing automated empathy because these typically disavow first-person perspective, self-reporting, and subjectivity. Neurophenomenology, in contrast, starts with recognition of individual brain uniqueness, it is oriented to a first-person perspective, and its interest is interacting with and preempting present-time consciousness. The phenomenological part is not simply an abstraction of the brain, but instead human experience plays a central role in scientific explanation. What commercial applications of neurotech are driving towards, then, is not just understanding of behaviour, connections, communication, content of a person's brain, or even the reconstruction of the neural factors underlying the experience, but also how experience impacts on material and functional dimensions of human being.

CONCLUSION

One role of technology ethics is to anticipate future harms and conundrums, which is not easy when the future is not written. The task is to grasp the technologies, their underlying principles, normative angles, nature of choices, social trends, intersecting interest groups, applied considerations, and to make a judgement call on the implications of given technologies and how proposed usage tallies against what is understood to be good and fair. Many claims made about neurotech stem from lab-based possibilities, often involving existing large fMRI scanners and/or smaller systems that are at very early stages of development. That our thoughts will be imaged by airport border security wielding hand-scanners is currently science fiction, as is the idea that healthy citizens would have part of their skull removed to ditch the PlayStation paddle. That brain activity may be more readily used to directly control objects and communicate is more realistic. The test will be whether citizens find value in the premise of touchless interaction with technology and media. What this portends is another question and while one may find ready vectors of critique in extractivism, cognitive liberty and freedom of thought interests, this chapter ends by flagging mental integrity concerns of neurophenomenology (material interest in first-person outlooks) and ongoing work on readiness potential (preemption of thought), that both portend unique ethical challenges for society.

Selling Emotions

Moral Limits of Intimate Data Markets

This chapter is a result of two intersecting trends: use of automated empathy to gauge emotion and expressions of subjective disposition; and
increasing interest in the idea that people should be able to better manage,
share, and potentially even sell personal data. The context is historic exploitation of personal data, where one's personal data, even about emotion
and mental disposition, is necessarily and forever subject to the whims of
globalised technology firms. A proposed solution from parts of the technology industry, and some governments, is that people should have greater
control over their personal data (Chakravorti et al. 2021). One approach to
enabling this is through apps that allow people to manage, share, and potentially sell data about themselves, involving nano-payments as a means
of empowerment and exercising commercial rights (Lanier 2013; Charitsis
et al. 2018). This idea is not new, with 'infomediaries' having origins in the
dot.com boom of the late 1990s/early 2000s (Hagel and Singer 1999). Still,
the idea that people may control and even be paid directly for their data
has stuck around. Today this is proposed to take place through Personal
Data Stores (PDS), also known as Personal Information Management
Systems (PIMS), or consent intermediaries, that all propose to assist how
people manage consent decisions through one interface (Lehtiniemi and
Kortesniemi 2017). The basic idea is that the PDS grants a person high

Automating Empathy. Andrew McStay, Oxford University Press. © Oxford University Press 2024.
DOI: 10.1093/oso/9780197615546.003.0010

levels of control over their personal data and ability to grant permission to access parts of the store of data.

A market-based solution to moral questions of informational self-determination, PDS proponents claim that if adopted at scale they would return power to individuals through increased choice, ability to reuse personal data for their own purposes across different services, and for them not to be beholden to platforms. This aligns with philosophies of technology based on architectures of decentralisation where, as per original social hopes for the Internet, data ownership and therefore power is not tied to platforms. Data in theory is held by users, who provide, share, rent, or sell their data to organisations. As a data intermediary, PDS would in principle give agency to people, regarding who sees and accesses their personal data, potentially about emotions, and on what terms. With hybrid ethics, a pillar of this book, involving questions about the terms by which we engage with technologies and services, PDS raise questions about agency, volition, whether the prism of market expansion is the best solution to data exploitation, but also whether certain options and terms should be denied be default to people.

There are ethical problems out of the gate. Risk is that decentralisation may give way to the emergence of new modes of plutocratic centralisation, a problem amplified by distributed and blockchain-based systems that obscure oversight and regulation. Indeed, criticisms of PDS are often based on the ideology of the service: here, design of PDS services and their promotion promise freedom from exploitation but are in truth another instance of capitalist apparatus (Mechant et al. 2021). This is supported by observation that these systems are predicated on 'engaged visibility' over 'technical anonymity' (Draper 2019: 188), meaning that privacy is less about removal of data from the 'market' than rerouting it. There is also the question of the overlap between commodity logics and human rights in that seeing one's own personal data as an 'asset' is awkward at best and immoral at worst. Moral disquiet is captured by Michael Sandel's (2012) 'Moral Limits of Markets' argument, that posits certain aspects of human life should be beyond monetisation and exchange. For Sandel, when utility and welfare maximisation through markets is uncritically applied to uses of the body (such as organ sales, for example), this raises questions about what should be in scope for markets. Does this apply to personal and biometric data, especially the sort generated through automated empathy? 'Yes!' is an understandable response, but does this reaction apply to every conceivable context? To explore these issues, the chapter outlines the nature of personal data stores, international and national governmental interest in them, and citizen views on the matter. With these insights in mind, it addresses the moral dilemma of whether data about emotion should be beyond the reach of markets and the logics of property.

PROPERTY REFORM THROUGH PERSONAL DATA STORES

There is no innate reason why technology platforms should maintain their control over online life, nor why law alone should be solely responsible for remedying this. Many regions of the world might argue that law has failed citizens from data exploitation. One response has been the suggestion of data cooperatives, or 'the voluntary collaborative pooling by individuals of their personal data for the benefit of the membership of the group or community' (Pentland and Hardjono 2020). Otherwise known as a data trust, Ruhaak (2019) defines this as 'a structure whereby data is placed under the control of a board of trustees with a fiduciary responsibility to look after the interests of the beneficiaries—you, me, society.' Lawrence (2016) likewise proposes these as a means of 'data governance by the people, for the people and with their consent,' likening current data-sharing arrangements to feudalism (serfs, labouring, lords, land, and exploitation). Although individual choice would be reduced, there are benefits from scale. Akin to unionised labour, they would have negotiation power due to the large numbers of individuals seeking to improve terms and conditions with powerful organisations (Element AI and Nesta 2019). The intention is to move from social and technical principles of centralised extraction to one based on a distributive model, an approach that embraces libertarian sentiment from the political right as well as left. Viljoen (2020) for example sees data as a collective resource for the common good, arguing that 'Such proposals view data not as an expression of an inner self subject to private ordering and the individual will, but as a collective resource subject to democratic ordering,' a point that could be re-articulated in terms of ethics work of Watsuji Tetsurō and the Kyoto School from Chapter 5 that foregrounds communal and interconnected aspects of human being.

If data cooperatives can be said to embrace a left and egalitarian libertarianism (i.e., mistrustful of concentration of political and economic power), suggesting collectively minded solutions to a perceived failure of lawmakers; PDS are similar but different in that they seek to decentralise power not to groups, but individuals, giving them a right-wing libertarian character. As a privacy enhancing technology, PDS allow people to specify the terms by which each element of their personal data can be used, for what purposes, and for specified time periods.

Market-friendly, PDS have four functional principles. First, through easy-to-use interfaces, they help their users store, download, carry, and port their personal data. Second, via meaningful and informed consent, individuals are given control over how their digital identities and personal data are used, allowing users to choose what data they share with selected individuals and organisations, and for what purposes. Third, partial revenue created by processing personal data may be given to individuals who provided that data. Fourth, while various nascent revenue models are being developed (Janssen et al. 2020), first-party PDS providers are typically paid by third-party companies seeking to access that data. In return PDS providers assure third parties receive (a) legal consent and compliance from app users, and (b) that users are happy rather than resigned to interact with third parties (thereby raising the economic value of interested users).

As well as increasing users' privacy and control over their personal data, some PDS providers also provide the ability to 'transact' (to sell, or otherwise monetise) their personal data (Janssen et al. 2020) via a more distributive economic model based on data dividends. These are small payments made to users and the PDS provider to access specific parts of users' digital identity. Deeply libertarian, the Data Dividend Project (2021) is among the better-known initiatives, alerting Americans to the fact that, 'Big tech and Data Brokers are making billions off your data per year. They track and monetise your every move online. Without giving you a dime.' Thus whereas extractivist theorists see the primary resource of the big data industries as people themselves (Sadowski 2019; Gregory and Sadowski 2021), requiring solutions of wholesale structural reform, PDS suggest different by pertaining to put users in control over of their personal data (including their biometric and emotion data).

A longer-standing PDS, *Digi.me*, centralises a person's social, medical, financial, health, fitness, music, and entertainment data. The app also permits sharing of inferences about emotion through one of its bundled apps ('Happy, Not Happy'), which allows users to track emotion patterns of their social media posts. Notably too, Digi.me also pulls biometric data from wearables, including Fitbit and Garmin, both having mood and emotional wellbeing trackers. Another of Digi.me's 'trusted' apps is UBDI, where users are excitedly promised they may 'participate in research studies and earn cash and UBDI when you choose to share opinions and anonymous insights from your data with non-profits, academic institutions or companies!' (Digi.me 2022). PDS are attracting significant commercial investment, with notable PDS platforms under development including the SOLID platform (stemming from 'social linked data') that is led by Sir Tim Berners-Lee (Web coinventor and Director of the World Wide Web Consortium) and developed and commercialised through the start-up, Inrupt. Other start-ups include Databox, MyDex, Meeco.me, Cozy Cloud, CitizenMe, and MyData. Industry-led political and ethical initiatives are exemplified by the aforementioned US-based Data Dividend Project that focuses on creating small amounts of revenue for Americans in exchange for their data. The Data Dividend Project and similar projects are of the view that people have a right to a share of profit made by means of their personal data, seeking to create data-as-property rights under privacy laws to allow Americans to be paid for their data.

Making the Strange Familiar

PDS are not new, but they are unfamiliar to lay users of the Web. Charitsis et al. (2018) for example notes start-ups such as Datacoup, Datum, DataWallet, and Powr of You that allow people to assemble social and bio-metric data from different sources into one place and be paid if they choose to sell. As with all platforms, scale, and network effects matter, particularly so if established platforms are being challenged. As above, the SOLID plat-form is interesting in this regard: it is an open-source personal data store developed by Sir Tim Berners-Lee. With a glaring problem being absence of a technical, political, and commercial infrastructure to recircuit data flows, the input of Berners-Lee makes the idea of ubiquitous decentralised data management a possibility. SOLID allows people to grant permis-sion to external apps and organisations to read and write data to 'pods' (namely, the store of data), returning control of personal data back to the user of that app. A key feature is that the pods compartmentalise data, re-ducing chance of third parties from interacting with data inappropriately (De Bot and Haegemans 2021; Mechant et al. 2021). Data is stored in an interoperable format, meaning that data is portable and can be used by dif-ferent systems and services.

SOLID is also notable because the British Broadcasting Corporation (BBC) is developing services based on the SOLID platform. With SOLID, the BBC's goal is to create a public service data ecosystem. This is signif-icant because the BBC has a public service remit, meaning that it must account for all levels of technical competence in the design of their serv-ices. This involves consideration of competence, usability, feasibility of how PDS would be accepted (or not), and how they would fit into eve-ryday life. With technology acceptance involving matters of usefulness and ease of use (Davis 1985) these are critical considerations given what is proposed: agency over data and partial recircuiting of data capitalism. The promise is big: to reprogram functioning and civic attitudes towards per-sonal data so that citizens have, and perceive themselves to have, greater agency over their personal data, through an interface (typically, a smart-phone PDS app) that is simple to use (Sharp 2021). If successful, the BBC will help normalise for lay people what currently is a difficult to grasp idea: managing one's own stores of personal data and having a greater role in what happens to it. Given that stripping of power from centralised platforms will entail some degree of data responsibility, design is key.

Beyond ability to have control over data, the goal is to enable a person to have extensive awareness of the streams of personal data they generate and to choose to share (or not) with diverse organisations. PDS apps may pull data from every app and service a person interacts with, certainly including sensors in phones (accelerometers, gyroscopes, global positioning, proximity awareness, magnometers, barometers, light-based sensors, face recognition, and biometrics). They would also represent data from wearable devices and apps, to generate a granular representation of a person. For example, a wrist-based wearable alone collects data about galvanic skin response (used to gauge emotion strength and arousal), electromyography (for nerve stimulation), blood volume pulse (heart rate tracking), skin temperature (which may signal stress or excitement), electrocardiograms (electrical signals of the heart), and respiration rate (that is associated with changes in affective states). Those interested in this data already include the insurance industry to gauge macro-societal patterns and individual risk, public and private health providers, firms reliant on geo-location targeting (not least the advertising industry), and retailers interested in knowing more about in-store reactivity.

Investigating 'Should We'

Both the UK's Department for Digital, Culture, Media & Sport, and the Department for Science, Innovation & Technology, lead the UK's National Data Strategy, which is dedicated to creating personal and organisational value from personal data (DCMS 2019). The initiative also champions the principle of data intermediaries, seeing scope for citizens to import 'personal data from providers such as social media companies, banks, hospitals, and the government, among others', and 'granting or revoking access to organisations such as GPs, banks, and online retailers, among others'

(CDEI 2021b). Similarly, the EU's General Data Protection Regulation (GDPR) was conceived in part to create a digital single market enabled by transparency, rights of access, data portability through interoperability, and allowing users greater control over their data. In principle, PDS do exactly this. Against the backdrop of a societal trust deficit in how personal data is used, PDS are held by the European Data Protection Supervisor to be a highly practical and human-centric way of delivering on the GDPR's promises of data rights while enabling new business models (EDPS 2020). The EU's proposed Data Governance Act (2020) is also significant in that it seeks to enable the reuse of certain categories of protected public-sector data and promote 'data altruism' across the EU through personal data sharing intermediaries (such as a PDS). In part about manufacturing trust and agency, the hope, at least in the UK and Europe, is that if citizens have greater control over their data, they will volunteer it for prosocial uses. Value extraction in this context might mean data for health or urban improvements, and encouragement of altruistic uses of data, donating for good causes. Although it is part of a political imitative to strengthen the single market for data in Europe, notably, it is 'against renumeration in any form' (European Commission 2020a: §1, 1). Instead, it takes its inspiration from research where data should be findable and reusable. The proposal flags that it seeks to move from the status quo of 'integrated tech platforms' to one based on 'neutral data intermediaries' (European Commission 2020a: 6). The latter favours 'common data spaces' although emphasising that 'natural and legal persons' should be 'in control of the data they generate' (Ibid.) through both sharing and pooling of anonymised data. Although it foregrounds public issues (such as climate change, mobility, and official statistics public services) and scientific research, it also includes privately funded research and the somewhat vague label, 'purposes of general interest' (Ibid.: 20).

While PDS are regarded by policymakers as a way of generating personal, social, and other organisational value from personal data, given the granularity of control over personal data that they afford users (Sharp 2021), the social history of PDS is deeply liberal, if not libertarian in the Robert Nozick tradition (2002 [1974]). This outlook emphasises values

of autonomy, minimal state size, entrepreneurialism, resolution of social problems through technological solutionism, free markets, economic
growth, and free choice at all costs. In relation to PDS this is dressed in
political and technological Web 3 language of decentralisation where 'it
is the internet itself that offers the services, and not a few big commercial parties' that provide solutions to surveillance capitalism (Mechant
et al. 2021), but the result is the same. Although the 1990s and early 2000s
language of cyber—and techno-libertarianism is often prosocial in language (especially when foregrounding agency over corporatism), there
is a lurking meritocratic worldview due to requisite levels of knowledge,
expertise, time; or inclination to manage apps, data, and terms of access.
This is amplified if people are paid for data (however small the payments
may be) due to shift of responsibility to what in effect is a microbusiness
of the self.

For good and worse PDS are about distributed ownership and responsibility, rather than a collective approach to the common good. The initial
attractions of self-freedom and liberalisation quickly bump into the collectively minded 'should we' question, in that while people should under-
standably be at liberty to choose to do what they like with their data and
property, what of the broader public good if people are able to sell digital
bits and insights about themselves? Answers to this can be divided into
two categories: (1) where we believe that PDS pertain to act in a person's
best interests; and (2) where we are more critical and do not believe this.

True Believers
Even if we agree that PDS developers are prosocial free-choice believers
who do not seek to take advantage of their users, there are still outstanding
ethical questions, especially given that universal human rights, such as
privacy (physical and digital), should not be contingent on financial income (OHCHR 2021). The risk is that a market-based approach to privacy is one that encourages unequal application of human rights due to
pre-existing different levels of financial incomes (creating privacy 'haves'
and 'have nots') (Spiekermann et al. 2015). Moreover, do multiple wrongs
add together to make a moral right? That is, does the existence of a data

industry legitimise its expansion? Expansion for believers would be based on principles of informational self-determination and human agency over decisions and property. PDS and 'data dividends' involve substantive political questions. Such services (and the wider premise of data dividends) regard data privacy through the prism of property, ownership, and licensing of data usage.

The view that privacy has origins in property rights has strong roots in the USA. For instance, DeCew (1997) (drawing on Thomas Scanlon, Jeffrey Reiman, and William Parent) argues that privacy is simply a late second-order expression of property rights (also see McStay 2014). Yet, in context of data dividends and micropayments through PDS, this has every chance of quickly getting complex. Even taking the argument of PDS proponents at face value, where people would weigh up the cost of disclosing information about themselves against a financial benefit (Beauvisage and Mellet 2020), this assumes that people are sole owners of data. As Draper (2019) also asks, what of when disclosures contain details about people who have not consented to release of that data? Urquhart et al. (2018) for example point to a common example on social media where data is not associated with just one person. For instance, a photograph taken of a group of people is not necessarily just one person's data, raising questions of consent, value distribution, and overlaps with data protection interests.

Even if PDS were agreed to be a viable privacy solution, disentangling who owns what would be hard. A property-based view would cast privacy (a human right) even deeper into market and commodity logics. Michael Sandel (2012) is useful here, arguing that when utility and welfare maximisation through markets is uncritically applied to uses of the body (such as organ sales and prostitution), this raises questions about what should be in scope for markets. Applied to the specifics of this book (biometrically inferred subjectivity), there is a category error: phenomena such as emotions should not be data-things to be bought and sold. As per the mental integrity interest throughout this book, because emotions are so core to human experience, decision-making, and communication, there is good argument that it is intrinsically right that datafied emotions (even the hyperreal sort) are kept away from domains of exchange.

SUSPICION OF MISALIGNED INCENTIVES

There is also the risk of what part 23 of the preamble to the EU's proposed Data Governance Act identifies as 'misaligned incentives,' where people are encouraged to make more data available than is in their best interest (European Commission 2020a: 17). Whereas free-choice believers see PDS and data trading as an expression of privacy rights for the public good, other motives are less prosocial. As apps are often free to use and because PDS services are paid by third parties, there is a clear risk of a PDS service simply being another extractivist actor, using questionable app design practices to nudge people to share and sell data, bucking principles of voluntariness. The language of extraction is not rhetorical but sits at the centre of UK governmental policy as, in language reminiscent of mining of material resources, it seeks to 'unlock the value of data' (CDEI 2021b). In general, PDS services seek to increase the amount of personal data in circulation, rather than lessen it, in the belief that all (user, intermediary, third-party, nature of service) benefit from richer data about people being in circulation (Lehtiniemi 2017). Other grounds for suspicion and criticism include observation that PDS miss the point, seeing that surveillance capitalism does not only exploit 'audiences', but creates 'worlds' that create audiences (Charitsis et al. 2018). Marxist in origin, this rests on the 'audience-as-commodity' thesis (Smythe 1981). Here, Internet users perform digital labour as they swipe, click, post, and contribute, because they generate value for others in the process, especially as insights derived from this data facilitates advertising based on this behaviour (McStay 2011; Fuchs 2013). Focusing on social media, Charitsis et al. (2018) also suggest that payments for data would be relatively low, making them akin to loyalty cards so people would become more loyal customers of platforms. Despite the appearance of empowerment, the existence of PDS would serve as means to close-off other more structural solutions to privacy. PDS and user micropayments from this perspective represent capitulation to surveillance capitalism, not a solution. Furthermore, numerous start-ups have tried and failed with this business idea. For example, distilling the pitches of five start-ups, Beauvisage and Mellet (2020: 7) see a trend: observation of platform profits, start-ups acting as brokers on subscriber behalf, redirection of profit, and emphasis on consent. The key problem is that without structural change (such as through law) even well-meaning PDS services from start-ups do not work, and highly arguably cannot, because start-ups are powerless in face of the platforms. The net consequence is that the PDS user is simply tracked by one more application.

SELLING EMOTIONS

Sandel's (2012) philosophical investigations into the moral limits of markets distil to the question of what should and should not be for sale. It is not a rejection of buying and selling per se, therefore not a rejection of trade. However, fundamental freedoms according to Sandel should not be expressed through the language of exchange. On the moral limits of markets, two criteria for Sandel are fairness and corruption. On his first criterion, *fairness*, the key question is to what extent a person (or group of people) is free to choose, or being coerced? This rests on the belief that it is unjust for people to buy and sell things under conditions of inequality and dire economic necessity (2012: 111). In a digital context, this is supported by data fairness observations regarding inadequate personal data literacy (Brunton and Nissenbaum 2015) and US-focused research that finds that most people skip reading privacy policies and consent material (Obar and Oeldorf-Hirsch 2020), making it difficult to see how people would engage with decisions about PDS data if subjected to repeated prompts for decisions. There is also the very practical burden of having to decide what to do with PDS data, and that lots of data requests from different services would make these decisions both many and difficult (Steedman et al. 2020). Difficulty for example takes the form of having to judge how one feels about given data and how a service proposes to use it (Pangrazio and Selwyn 2019). In context of automated empathy this would expand the question from willingness to share information about web traces and websites visited with third parties, to interest in sentiment and biometric data that is arguably more intimate in nature. Overall, this raises the question of whether PDS offer an illusory and deceptive sense of control, if making sense of the consequences of data disclosure decisions does not become easier than it currently is (Lehtiniemi and Kortesniemi 2017). Liberal principles of freedom, control and agency begin to appear as a sales-pitch for systems built to serve new extractivist actors. For Draper (2019) a key problem is reliance on the classic liberal principle of autonomy and personal responsibility, pointing instead to feminist relational forms of autonomy that are grounded on collective solutions and obligations. This is again reminiscent of Chapter 5 on Japanese thought, the Kyoto School, the context imperative, and interest in accounting for the full costs of technologies. Here leading thinkers refused to see the individual and the collective as binary premise, foregrounding instead 'in-betweenness' to create harmony between individuals and communities.

It is questionable then whether market choices though PDS would be free due to potential asymmetries in PDS, including issues of literacy, limited control over designed interfaces, pressure through time burden, and pressure through repeated demands for decisions. Sandel's second criterion of *corruption* argues that certain things and services involve a degrading and objectifying view of humans and human being, meaning that exchanges cannot be resolved by appeals to fair bargaining conditions. This is an important but easy argument in cases of slavery and organ sales, but 'degrading and objectifying' are not analytically sharp terms. Limits gets hazy with less visceral cases, including emotions and data about emotions. However, the strength of the moral limits of markets argument also stems from *how* goods or practices are corrupted. Corruption in relation to markets occurs through a category error, where a thing that is not usually labelled as a commodity is categorised as a commodity.

Historically commodities are economic goods produced for a market, mostly through growing, mining, and manufacturing. The process of commoditisation occurs when it no longer matters who is supplying the goods to a market, assuming a quality threshold is met. Of course, personal data about interests, preferences, and psychological disposition inferred through online traces has long been commoditised under dubious terms and conditions (certainly failing the fairness criterion of freedom to choose). Emotion has also long been commoditised, with Autonomist

Marxists having long detailed the 'progress' of industrial commoditisation from things and outside physical environments, to include commodification of subjectivity (of self and in relation to others). This critical theory recognises the changing nature of production and financialisation, whereby 'emotional corporeality is subsumed and incorporated by the production of value' (Berardi 2009: 109). Others sympathetic to Autonomist Marxism point out 'affective economies' and the psychographics and emotion-inferencing of modern marketing (Andrejevic 2013). While coming from *very* different philosophical places regarding the social role of markets, Sandel and Autonomists would agree that amongst the critical politics and desire to break-up affective economies, emotion should not be categorised as a commodity. Criticism by category is helped by trying to square the idea of property in relation to emotion, which is awkward at best. Whether one defines emotion in autonomic, experiential, intentional, communicative, judgemental, historical, cultural, or social terms, none of these definitions of emotion readily align with the principle of ownership and property. The definitional category problem is amplified by links with emotion to fundamental freedoms, especially those involving mental integrity.

Hybridity and the Moral Limits of Emotion Markets

With hybrid ethics involving questions of what sorts of technologies do people want to live with, on what terms, how do they affect the existing order of things, and at what cost do the benefits of modern technologies come, are there then situations where selling data about emotions may be acceptable? The quick answer could be 'yes' as it is not emotion, it is just data. The problem is that the personal data in this context exists at all stages for a purpose, and it is about something, which here is emotion. What is at stake is people, more so than computer bits. Even if the reading of a person is partially or wholly wrong, it is still about them, potentially involving judgements that will have a material impact. One cannot argue that just because it is datafied emotion, it is divorced from a person's emotional life—even when people are misjudged. However, although we know there exists a category problem and that there are inherent corruptive mental integrity implications for emotion and other empathy data, this does not necessarily trump rights of choice and autonomy to sell. While distrust of PDS developer motives and market are reasonable, and accounts of individualist autonomy based on personal responsibility may be disliked, it requires a strong argument to deny people the ability to sell their data, especially when people in general may stick with the status quo.

Whether one is morally *obliged* to keep data about emotions out of range of markets depends on philosophical approaches to markets, liberty, free choice, but also what these principles mean in terms of PDS and data dividends. On liberty, the principle of PDS echoes John Stuart Mill's (1962 [1859]) belief that if a person is not harming others, and they have independence and rights over their own bodies and minds, they should be free to do what they choose. Libertarianism takes this further, speaking of ownership over the inviolate self who (echoing Kant) should not be used as tools, means, and ends of others. Deeply antipaternalistic and unwilling to prohibit activities in the best interests of others, this class of libertarianism argues for a reduced and 'minimal state'. What matters for libertarians such as Richard Nozick is choice, with him simplifying his philosophy via the maxim 'For each as they choose, to each as they are chosen' (2002 [1974]: p. 160). Nozick has an ally in Hayek (2001 [1944]), where individuals are also mostly allowed to follow their own values and preferences. Against interference and regulation, the Hayekian oeuvre sees economic life in terms of 'spontaneous and uncontrolled efforts of individuals' (Ibid.: 15), 'creative energies' (Ibid.: 246) and that 'unchaining of individuals' leads to 'the satisfaction of ever-widening ranges of desires' (Ibid.: 16–17). Likewise for Friedman, competitive capitalism, private enterprise, and free markets, are the ways to deliver a truly liberal society, *'provided the transaction is bi-laterally voluntary and informed'* (1962 [2002]: 13 [emphasis in original]).

Yet, while people say they want more control over their personal data (Hartman et al. 2020), are markets and property the best prism through which to address subjectivity, personal data, and the odd idea of selling emotion (strange because of the value category mix-up)? Deontics of human rights should be challenged from time to time, if only to affirm their worth. Beyond category error, the problem with libertarianism and selling kidneys is circumstantial coercion and the likelihood of terrible human need to sell an organ. Does the same principle apply, albeit less dramatically to emotions? Should we be concerned that high-value data about the body and emotions has scope to be sold by the financially vulnerable when they would otherwise not have chosen to do so? Again, concerns about choice, agency, voluntariness, consent, being informed, and privacy 'haves' and 'have nots' in data markets, comes into play.

Are There Any Situations When It Is OK to Sell Intimate Data?

There are few studies on public thoughts and feelings towards PDS in part because it is a relatively niche topic for social science and that researching new technologies with the lay public is challenging, due to their lack of familiarity. However, it is something asked for by the field, with De Bot and Haegemans (2021: 4) remarking that more insights from the public are needed. It is useful too, to pre-empt intersecting trends likely to intersect—increase in biometric profiling to try to measure subjectivity, and decentralised PDS that have scope to create new data markets. In a rare study of the terms by which UK citizens would share and sell data through a personal data store, at the Emotional AI Lab we considered (1) types of personal data from which inferences about emotions, moods and mental states might be inferred; and (2) types of organisations that a person might sell this data to (Bakir et al. 2021a, 2021b). While the UK is not representative, that attitudinal studies are rare on this topic perhaps lends interest in the results. Of 2,070 respondents to a demographically representative online survey, half (50%) of the UK adult population are not comfortable with the idea of selling data about emotions, moods, and mental states, but a large minority are comfortable with this premise (43%). Economically, we found that people most uncomfortable with the idea of selling biometric or emotion data had a low income of £14,001–£28,000, with the group below this (£14,000 or under) being only marginally less uncomfortable. Potentially, these two demographics are regularly targeted with emotionally manipulative content (for instance, for payday loans, payment plans, and consumer items that they cannot afford) and hence may be more sensitised to the manipulative uses to which emotion data may be put. (Qualitative work in this area would be useful to see if the speculative explanation has merit.)

Despite well-known imperfections of online surveys and asking lay citizens propositional questions about technologies they are unfamiliar with, we had two stand-out findings. The most important finding for moral limits of markets questions is the finding that in principle most citizens are *not* comfortable with the idea of selling data about their emotional life; but this is closely followed by the second finding that there is a clear division in the types of organisations a person is willing to sell to. This division has implications for the category-based argument and seeing emotion through the prism of commodity markets. As depicted in Table 10.1 (derived from Bakir et al. 2021b), we found that where citizens are willing to sell data, this was to health, charity, and academic research-based organisations. Where they are unwilling, is to organisations of a clearly commercial and political campaigning sorts. Interestingly and somewhat neatly, private drug and

Table 10.1 WILLINGNESS TO SELL EMOTION DATA TO DIFFERENT TYPES OF ORGANISATIONS

Organisation emotion data would be sold to	Willing to sell	Unwilling to sell
UK National Health Service (NHS)	65%	35%
Academic scientific research	61%	39%
Mental health charities	58%	42%
Private drug & health companies	51%	49%
Financial services	45%	55%
Advertisers & marketers	41%	59%
Social media & other tech firms	38%	62%
Political parties & campaigners during elections/referenda	37%	63%

SOURCE: Walnut Unlimited online survey, 29 September–1 October 2021, 2,070 respondents

health companies landed in the middle of not-for-profit and commercial and political uses of emotion data. Private drug and health companies are of course just that, commercial entities, but they also have a role in society beyond that of most other commercial organisations. Notably too this UK survey was conducted during the COVID-19 pandemic when commercial vaccines were becoming proven in their efficacy against prevalent strains, and when television coverage of private drug and health companies was highly positive.

This introduces new category dynamics because the assumption of moral limits of markets questioning and the 'commodity' category outlined above, is of technology companies and marketers who want to use emotion and empathy data in ways that are not obviously (if at all) prosocial. What is clear from the survey is that most UK citizens do not like the idea of a society where intimate states are commoditised by the types of organisations Autonomists and critical data scholars might think of in the first instance. Notably, UK citizens dislike use for political influence even more (echoing separate work by the Emotional AI Lab on datafied emotion and political influence). Use for prosocial research and care is different, perhaps requiring adjustment on acceptability of being paid for data about interior states. Purpose clearly plays a role in this regard and, although we did not ask citizens what their value assumptions were, it is reasonable to assume they are based on well-being, encouragement of flourishing, care, and use of data for advancement for mental and collective mental integrity.

From the point of view of the EU's 2020 proposed Data Governance Act and other regional interest in creating new forms of social value from data, this is interesting. UK citizens are willing to be paid and while the proposed Data Governance Act is 'against remuneration in any form' (European Commission 2020a: §1), one might speculate that renumeration is imagined in the category of profit-motivated business rather than payment for medical and academic studies. If PDS were used in Europe, and beyond, in this way, this should still raise ethical questions due to the ease and frictionless nature of participation via a PDS. Mental integrity and Sandel's corruption concerns are not fully addressed either in relation to less well-off people signing-up for potentially multiple studies they might not otherwise have chosen to participate in. Although there is a long precedent in payment for medical trials and academic studies, which are arguably beyond commodity exchange as we routinely understand it, the PDS would make it easy to participate in many studies, potentially totalling to an income stream for those with otherwise low incomes. This begins to look like something more market-based than the more dignifying act of compensating for participation in prosocial research. Any move to use PDS for research must explicitly deal with this point. Yet, overall, the finding is clear in that healthcare and the public good appears to matter to (British) citizens, as they indicate willingness to engage in incentives to share data about intimate dimensions of their lives. This is especially so as we found few demographic differences. For the most part, majorities are willing to sell their data in non-profit research-based contexts across both male and female genders, all ethnicities, all household incomes, all socio-economic groups, all ages under 55 years old, and most employment statuses.[1] As ever, policy should be informed by experts, especially given that non-expert survey respondents were tasked with answering questions about technologies they will have neither seen nor used. Nevertheless, the clear public/private divide among citizens (with private health and drug companies in the middle) may provide fertile ground for expert reflection.

CONCLUSION

The implicit proposal of personal data stores is to use decentralised principles to break the dominance of digital platforms that function through extraction of personal data. Of the principle of selling data about emotion and others intimate states generated by automated empathy profiling, this chapters recognises political interest in the potential of PDS as a means of both protecting privacy and simultaneously liberalising data markets. Neither data trusts nor PDS are new, but ethical, technical, usability, and literacy challenges have prevented serious rollout. For these reasons the British Broadcasting Corporation (BBC) development of services based on Tim Berners-Lee's SOLID platform is significant. As the BBC is a public service organisation, it must simplify complexity to render the service of interest to all UK citizens, rather than the digerati. Although the BBC are not investigating data about emotion or enabling data dividends, others PDS are (Data Dividend Project 2021).

Critics point out that rerouting portions of the financial value of personal data back to individuals that generated it represents expansion and capitulation to surveillance capitalism, not its solution. The reason why criticism of PDS is strong is not because of aversion to people receiving micropayments, but that this lure equates to neoliberal misdirection from an Internet and Web founded on the common good. This rests on belief and possibility of a publicly owned online environment, just as we have commonly owned public services and other civic infrastructure (Fuchs 2021). Yet, the principle of PDS as a means of managing personal data and identity do not only involve marketing and advertising, and systems that are broadly exploitative in nature, potentially including public services. Indeed, although surveys of technologies people are unfamiliar with are imperfect, the survey data presented in this chapter indicates an interest in use and sale of emotion data for the public good. Citizen views are backed by the observation that it is a value and category mistake to see data about emotions and mental states through the prism of property. Most UK citizens surveyed by the Emotional AI Lab appear to agree. When asked about the principle of selling data, they were warmer to the idea of organisations that they believe to serve the public good in relation to health. This provides a twist takes on what Sandel sees as corrupting market logics. The ethical question becomes one closer to medical ethics, that grapples with whether money is an undue inducement in research. Given political interest in deriving pro-social benefit from public data the affordances of PDS will have to be factored into medical ethics, given the ease by which a person might sign up for numerous studies. Properties of PDS apps thus risk turning compensation and renumeration into an income stream, a point of concern raised about privacy haves/have nots in relation to manipulative targeting on basis of emotion data. This chapter concludes that ongoing legal alertness to commodification of emotion and other data collected through automated empathy is required, rejecting the extractivist premise that because data about psychology and emotion is already commodified, this is moral license to create an empathic data market through PDS trading. Yet, use of PDS to share and sell data, and opt-in to studies is of interest, where it unequivocally (and independently assessed) promotes individual and collective mental integrity, although undue inducement of the financially poor will remain a concern. Given phone and wearable sensors, and scope for self-reporting, this chapter tentatively concludes the moral limits of intimate data markets is restricted to not-for-profits and prosocial uses, such as public health.

Uncertainty for Good, Inverting Automated Empathy

By means of use cases that are happening now, likely to happen soon, and further towards the social horizon, this book has sought to anticipate and identify problems with *automated empathy* systems before they are embedded in everyday life. Given breadth of potential uses and scope for ubiquity, upfront awareness is needed because preventative action will become increasingly difficult as such systems are established in diverse social domains. In surfacing an array of problems, this chapter seeks to address these by inverting the core concept of automated empathy to envision what good could look like in relation to the presence or absence of these technologies in everyday life.

RECAPPING THE CORE ARGUMENTS

Automated empathy systems sense, see, hear, and measure people, with a view to 'feeling-into' everyday life and reacting somehow. To an extent this is an easy theoretical move, one that connects the capabilities of modern technologies with psychological accounts of human empathy that are based on observation and theorisation of disposition, such as that of Alvin Goldman (2008). For people and technologies alike, this observation-based account of empathy involves surveying, measuring, remembering,

Automating Empathy. Andrew McStay, Oxford University Press. © Oxford University Press 2024.
DOI: 10.1093/oso/9780197615546.003.0011

and creating rules for subsequent engagement. Certainly cold, such a view of empathy lacks mentalistic, bodily and imaginative attributes, that would otherwise be associated with empathy. It is even further from sympathy, where a person is said to share the feelings of another (with *sym* meaning 'together' and *pathos* emotion or feeling).

There is also a broader observation being made in this book, one that sees simulative properties of automated empathy (or appearance of understanding) as becoming more prominent in devices, services, agents, and environments. Despite the problems documented in this book about automated empathy, not least emotion recognition and what may be unfixable problems with these systems, technologies will likely continue through a variety of means to get more personal with people. We should be wary of criticism based on oversimplicity, accuracy, bias, and reductionist understandings of physiognomy. It is not that these critiques are wrong but that the industrial response is already showing signs of parrying these critiques with arguments about the prosocial benefits of mass automated empathy and that accuracy may be improved by granular and more dynamic profiling of people. Multimodal and context-sensitive automated empathy systems, for example, will rely less on capturing frozen facial expressions, than assessing change in physiology in relation to personal baseline thresholds and known contexts to infer subjective disposition and qualitative experience. Multimodally inferred flow, fluctuation, continuity, and context become the key watchwords, which are quite different from static and universalised approaches long associated with emotion recognition technologies.

Although there is conceivable scope for some social good in these human-systems interactions, how these technologies have been put to work so far does not inspire much optimism. Rather, as Chapter 1 notes, there is every need for governance and policy bodies to recognise the negative potential of these technologies and closely regulate their use in sectors that have conceivable material impact on people. Attention is especially warranted where there is a lacuna in governance. Examples includes data protection questions concerning the processing of biometric data where a person may not be identifiable (as with transient data); and, separately, where nonbiometric proxies (such as vehicle telematics) are used to make judgements about human behaviour, emotion, and intention.

The emotion aspect of automated empathy has been much criticised, primarily because industrial emotion-sensing technologies are associated with academically, ethically, and politically questionable methods. One of my central arguments has been that industrialised emotion has a 'hyperreal' quality, drawing on media philosopher Jean Baudrillard (1993 [1976]) and his idea that metaphors may stand in for reality, and may be presented as real. Applied, I defined *hyperreal emotion* as industrial prescription of emotion that takes place in the vacuum of absence of agreement on what emotion is. This entails longstanding criticisms of accuracy and method, but also what Stark and Hoey (2020) phrase well as a 'prescriptive' account of emotion. I made a similar argument in *Emotional AI* (McStay 2018) albeit less concisely, highlighting that the problem with simplistic accounts of emotion is that these prescriptions (or 'articulations' as I put it) have power. This is especially so when supported by convincing and attractive visualisation of changing expressions and numeric scoring of emotion. One problem is that while supervised machine learning through basic emotions is simple and seductive, and to an extent even effective through better-than-chance success of labelling of emotion, they are nonetheless highly limited (Azari et al. 2020). The approach is selective in what it looks for, seeing only what it is told to see, missing much about the relational, contextual, and situational nature of emotion. The problem is less that it gets everything wrong, but that it misses so much regarding the richness and diversity of emotional life. Consequently, through a narrow and biologically reductionist worldview it *imposes a prescription of emotional life that has no ground truth, a hyperreal*. If it were not for increased interest in applying hyperreal emotion to industrial, educational, safety-critical, health, workplace, and public contexts, this would not be so problematic. However, as this book has shown, there is growing interest in widespread applications. As such, the limitations of hyperreal emotion must be recognised and addressed.

Connected, another key argument of the book is that of *unnatural mediation*, that signals concern about the political problems of mediation.

This derives from a longstanding tradition in media and communication studies that questions filters, processes, and the factors (human and technical) that inform how people and things are represented (Silverstone 2002). It sees that meaning is not simply transmitted through media, instead seeing multiple problems regarding how meaning (such as that of emotion and aspects of subjectivity) is defined, impacted upon, and produced by the automated empathy industry. Many societal issues were explored through the lens of *unnatural mediation*, some pressing and immediate, others future-facing but going to the heart of what it means to interact with automated empathy systems. Pressing societal issues, connecting with wider racial justice movements, are issues of racial bias in automated empathy: most specifically, whether emotional AI is simply remediation of old racist 'science' in the form of physiognomy (or more accurately for emotion, pathognomy). Other pressing *unnatural mediation* issues are environmental and planetary, specifically the exponential carbon emissions that arise from trying to increase the accuracy of automated empathy systems through force of computation rather than revision of psychological understanding. Also in need of immediate attention are the impacts of automated empathy systems on children and education, especially the need to sift between what are good and prosocial technologies and what are detrimental to child experience and mental integrity. Given the rapid consumer-facing deployment of automated empathy systems in cars, we also need to consider questions of safety, of whether privacy is contingent on identification, and whether society would benefit from cars and technologies that themselves 'feel' vulnerable. Already present in certain sectors is the impact of automated empathy systems in the workplace. Tackling corrosive uses in the workplace (especially the robotisation of Angela, the call centre worker who was forced to feel-into the disposition of workplace analytics), the chapter on the workplace considered whether automated empathy may serve union and collective labour efforts. Looking further towards the sociotechnical horizon, the book then examined brain-body-computer interfaces that pertain to understand and even preempt first-person intention. Final consideration was given to the moral merit of people selling data about their emotional and subjective lives, and the innate attraction and inherent problems in decentralised solutions to negative impacts of informational capitalism.

The next argument underpinning this book, helping to identify socially corrosive uses of automated empathy is that of *mental integrity*. Appropriated from the neuroethics literature (Lavazza 2018) but tweaked to have application beyond the brain, and expanded to encompass *collective* mental integrity, this argument is subtle. In part it is a catch-all label for privacy, security, agency, freedom of thought, and autonomy, which are obviously familiar rights and values. The more nuanced part is sensitivity of mental integrity critique to psychological aspects of relationships with technology, especially to first-person phenomenology and technologies that pertain to understand us inside-out. For a book about automated empathy, this is a vital point.

The next orienting argument of this book is that of *hybrid ethics*. Perhaps unlike other books on connected topics, this book did not reject technologies that function in relation to intimate data at the outset. Even at its conclusion and having identified deep flaws in method and application, it still does not. With more than a nod to aspects of writing from Bruno Latour, Peter-Paul Verbeek, and Sherry Turkle, the hybrid stance asks: (1) what sorts of technologies do people want to live with; (2) on what terms; (3) how do they affect the existing order of things; (4) at what cost do the benefits of given technologies come; and (5) how can the relationship be better? Its ethical starting point is that people are increasingly continuous with their technologies and entangled in data-rich environments. The risk of this straightforward observation is accusation of submission to contemporary human-capital-technical arrangements that function through dubious mechanisms. The goal throughout, however, has always been about better human-system relationships, as well as severing where appropriate. In addition to the five points above, the book was also motivated by a *context imperative*, a heuristic that demands that we consider the full costs (and benefits) of any proposed technology and its uses. Inspired by climate and related environmental dimensions, it stimulates the will to address sustainability concerns in relation to spiritual, psychological, biological, social, cultural, economic, legal, and political dimensions of automated empathy outcomes.

KEEPING IT IN THE COMMUNITY

Having recapped the core themes of this book, this final section of the book considers the underpinning principles of what good might look like for automated empathy. Technologies that function in relation to intimate dimensions of human life have been critiqued on grounds of scientific racism, innate bias, hyperreality, effectiveness, pseudoscience, privacy, freedom of thought, impact on human dignity, and individual and collective mental integrity. There are then many concerns to be addressed, but it is also useful to acknowledge what good could look like in relation to the presence or absence of these technologies in everyday life, if only to help highlight what is problematic with current arrangements. To go further, this book does not see the premise of technological empathy as innately problematic and, with genuine care and very large caveats, there may be positive uses. Broad examples involve making technologies easier to use, new forms of interactive media content, novel modes of expression and communication, aesthetic experience, understanding how people feel about issues, improvement of therapeutics, and promotion of health and well-being. Given that the lifetime of a book is one of years and hopefully decades, there is arguably little merit in detailing the technical specifics of any proposed 'good' application, due to uncertainty of what technologies will feature in coming years. However, what is highly likely is that relationships both with, and through, technologies *will* become more intimate. The final few sections of this book suggest principles of interaction with automated empathy. The advantage of this is that one can take advantage of the vector-like properties of principles, where ideas about what good might look like may be scaled into unclear future technical contexts without loss of applicability.

Flipped Empathy

As the book has shown, commercial and other organisational usages of automated empathy show few (if any) unequivocally prosocial uses, and many signs of being exploitative. Key fail-points of automated empathy are potential for judgement to negatively impact on a person in some way, and absence of scope for co-construction. The latter represents the extent to which a person may interrupt, override, modify the behaviour of the other, and, in a meaningful sense, be in control of the relationship. This is reminiscent of urging of attention to 'small data' rather than big data. A small view attempts to renew relationships with data; that is, to recollect that data in the case of people are entirely about their bodies, family members, friends, subjective experiences, likes, choices, achievements, efforts, and failures (Lupton 2018). Relationships with technologies is key and if the language of empathy is to be at all sensible in connection with technology, people must be seen as dominant partners who are able to clarify and correct understandings within their relationships with a system. What good looks like for empathic systems is where people can engage with their bodies and communications to facilitate novel forms of reflection, therapeutics, learning, creativity, and expression. For this to happen, the entire premise of automated empathy needs to be flipped. This involves the following inversions:

As should be clear by now, there is every need to be suspicious of uniform affects, universalised emotion, and emotion fingerprints (Barrett 2017). At heart of the inversion taxonomy above is impetus for a revival of longstanding principles of interactivity (Lister et al. 2003), an arrangement in which people may intervene and be active in generation of *their* meaning. This situated view is hopefully by now beginning to seem self-evidently necessary due to problems inherent in universalised accounts of emotion and interaction with human disposition (empathy). With interactivity being the question of real-time control over computational processes (Suchman 2007) it is also the extent to which a person may interrupt, override, and modify a computational operation or system. As we have seen throughout the book, the lack of ability of people to actively

Table 11.1 FLIPPED AND INVERTED EMPATHIC SYSTEMS

Standard	Flipped
Universal	Local
Big tech	Small tech
Homogeneous	Heterogenous
Deductive interpretation	Inductive creation
Small number of classifiers	Large palette of emotion
Colonise	Indigenous expression
Passive	Active
Closed system	Modifiable
Prescribed labelling	Self-labelling
Static	Emergent
Cloud	Edge/local/non-networked processing
Marginalising	Inclusive
Surveillance	Sousveillance
Unnecessary collection	Purposes limited
Control and management	Reflection and self-learning
Tool of others	Self/collective-determination
Managerial	Worker-first
Monopoly	Decentralised

input into systems is keenly notable due to what are often hyperreal universalising logics. This represents a loss but also an ethical opportunity for automated empathic technologies in that a prosocial deployment is one that respects scope to input, modify, and override where necessary. This is a key argument of this book: both methodologically and ethically, a more local approach is needed to automated empathy that allows people to configure and modify their relationships with such systems.

Feminist epistemology is useful in this regard as it is often based on critique of flawed conceptions of knowledge, knowers, objectivity, and scientific methodology (Anderson 2020). With the nature of emotion being so problematic, a situated approach is not only prosocial but an intellectual 'must-do'. If the language of empathy is at all sensible in connection with technology, people must be seen as partners who are able to clarify and correct understandings within their relationships with a system. In context of empathic technologies, this may be to further all sorts of interests, such as shared intimacy, communication, games, self-mood tracking, health, aesthetics, and so on. Indeed, the longstanding connection between emotion, empathy, expression, experience, art, and other means of creative production, might mean that datafied inferences about emotion need not be innately bad. Although it is routine to critique the exploitative nature of modern media technologies, it is important not to lose sight of the fact that nonexploitative arrangements are neither utopian fantasy nor capitulation to Big Tech. The goal rather is to reclaim a technological environment that runs counter to centralisation and gross organisational dominance.

The idea of flipping empathic systems should even be familiar, especially in reference to conceptions of interactivity from active audience ideas about media usage, that in turn are based on interpretive acts, local meaning-making, the specifics of circumstances, and how encounters with media and data may affect both the person/group, and the system itself. There is also a political dimension to enhanced interactivity that has roots in ambition to democratise media and related technological systems. This eschews one-way information flows, preferring reciprocal, localised, and relationship-based approaches. Technologically, a bottom-up approach has scope to take advantage of proliferation of edge computing and IoT devices, associated with low power rates and small systems with relatively few parameters and layers. A device might, for example, recognise activity either on the device itself (such as orientation sensors) or through sensing a body, allowing for self/group-created labels to register affect and related episodes in known and novel contexts, and annotation and correction of labels for those contexts. These reduce, if not remove need, for engagement with cloud-based services, decreasing energy costs, and increasing privacy.

Self-Annotation and Play

Self-labelling, or scope for self-annotation to create local meaning, is certainly smaller in scope than the universalising systems of Microsoft, Amazon, Meta, and others. Small and diverse is good in this regard, shifting focus from homogenous hyperreal representation to one based on heterogeneity, this involving localised experiences of self, others, systems, and place. As a starting point this is one based on a dimensional account of emotion that foregrounds valance and arousal but does not impose prescriptivist and hyperreal emotion labels, allowing instead for self and co-created accounts. As Ott (2020) puts it in her challenging of Spinoza and Deleuze-inspired accounts of affect; historical, local, contextual, and temporary accounts of affect are preferable, to allow for small and home-grown fields of relations. While Ott's purpose was an intellectual counter to the universalising and homogenising cultural theory of affect, her point is relevant in that she seeks a vernacular of ethnocentric affective articulations that escapes assigning of identity and experience. Such a flipped and bottom-up conception applies well here, allowing for micro- and subcultural labelling. The idea is to avoid hyperreal prescriptivism, allow people to label what input data means, and to co-create with intimate others a lexicon of emotions and communication. In effect this is a high-level proposal for what might be phrased as *sensual autonomous zones*, closed groups that allow for play with sensors and affect labelling, intragroup defined affective vernacular, expansion of communication bandwidth, and a deeply heterogenous (and flipped) stance to technologically mediated empathy. Indeed, in an early critique of emotion labelling systems, Boehner et al. suggest that 'breaking out of the closed world of codification is to shift primary focus from the system's internal representations to users' interpretations and experiences around the system' (2008: 5). Now with experience from seeing the emergence of automated empathy across multiple social sites (such as education, transport, and work, for a few examples), the need for inversion and pro-social small tech takes on extra impetus.

The point above about play and the overall hybrid approach that seeks to improve terms of interaction with automated empathy can be expanded. Philosophies of play point out that play is voluntary, cannot be forced, is not a task, is done at leisure, involves separation from everyday life, has a defined duration, and is predicated on freedom (Huizinga 1955 [1938]). Also, while one might associate playfulness with openness and absence of rules or boundaries, the opposite is true. Although it involves uncertainty of outcome and pleasure, play is also about contextually contingent rules, and separation from wider goings-on in life, closed imaginary universes, and communication that sits outside the normal order of things (Callois 2001 [1958]). Play, then, has a surprisingly precise and bounded meaning. It should also be a *safe* setting for creative endeavour, experimentation, expression, strengthening, and learning. With play being about separation of domains, it is perhaps useful to repurpose Huizinga's (1955 [1938]) requirements of play for datafied play in context of sensual autonomous zones. Voluntariness *must* be respected, and governance care (technical as well as legal) should be given to ensure that data (personal and group data) remains bounded within the play context. This is not straightforward as settings will be saved and memory will be used, but the goal must be to ensure that data does not spill from safe autonomous play domains.

Participation

An active, playful and participatory approach also allows individuals and communities to build alternative infrastructures, something that again has political dimensions due to 'design justice' (Costanza-Chock 2020) and need for recognition of who may be marginalised by universal and ubiquitous systems. Although involving political redress, there is a creative and expressive dimension to sensual autonomous zones. Exploration might be made of Birhane's (2021a) putting of soul, stories, interest, motivations, and participation at the heart of sense-making with data, and knowledge that admits of concrete particularities and affective connections. This stands in contrast to hyperreal buckets and universal labels. Birhane's

inspiration is Afrofeminist, reflecting grounded analysis and action in relation to histories of colonialism, racial formation, and gender hierarchy of
the various European nation-states in which Black women live (Emejulu
and Sobande 2019). From the rights and needs of marginalised groups
there is a clear lesson in the value of ground-up participatory approaches,
localism, richness of in-group communication, self-annotation, and use
of technologies to tell stories and better understand self and others.

This flipped and participatory understanding is potentially creative,
allowing for user input, modification, self-annotation, reflection, and
safe self-learning about emotions and experience. In contrast to self-
examination through eighteenth century physiognomy handbooks, or
modern industrial-scale computational inferencing, self- and community-
based annotation does not provide prescriptivist and hyperreal answers.
It facilitates, enabling novel ways of creating meaning, autobiographical
expression, co-creating and interacting with others, labelling, emotion
language creation/play, and therapeutic understanding of self and others.
Words, pictures, bespoke emoji, haptics, and similar, are all means of
rendering biometrics for personal and interpersonal co-creative use, perhaps for self-learning (through patterns over time), but also as a means
of sharing and continuing ambient awareness through media. Mediated
empathy in this context is explicitly based on co-presence and being-
with, and shared experience. This involves extended affective experience
by means of autonomous zones between intimate others (partner, lover,
child, parent, friend, and so on). Indeed, inverted empathy systems might
allow intimate groups to design looks, sounds and haptics languages, potentially for mixed as well as physical reality contexts. This would serve to
counter the problem of prescriptive labelling and create ways of sharing
and querying how others are doing. Such an approach bucks the hyperreal and moves towards *natural mediation*, one based on relationships and
agreed truths of communication and meaning.

Broadly, these principles and observations of what good looks like is
a humanities-based appeal to reimagine automated empathy as less prescriptive, and to admit of plurality, local experience, indigenous expression, and sensual autonomous zones. The balance of power should be

playfully weighted towards citizens. This is to: (1) facilitate greater interpretive license of physiology; (2) keep data local and under control; and
(3) enable self-labelling, local meaning-making, and co-creative uses. An
inversion of psycho-physiological systems from the universal would allow
different conceptions of emotion to be explored, novel self-programmed
self/group-understanding, local languages of affect-based communication, and new palettes for creative and therapeutic endeavour. This is
smaller, but with care it can be good.

CHAPTER 1

1. With this book written during the period the AI Act was being debated and passed, so still subject to changes, readers are encouraged to consult the final version of the Act and its Annexes that provide detail on how given applications and use cases are being regulated.

CHAPTER 3

1. In Western thought this has diverse roots in a wide range of non-Germanic philosophers, including Frances Hutchinson, Lord Ashley Shaftsbury, Herbert Spencer, Charles Darwin, Adam Smith, and David Hume, among others.

CHAPTER 4

1. See:https://books.google.com/ngrams/graph?content=ethics&year_start=1800&year_end=2019&corpus=26&smoothing=3&direct_url=t1%3B%2Cethics%3B%2Cc0. A link and observation made in a Twitter exchange by Gianclaudio Malgieri, Associate Professor at in Law & Technology EDHEC Business School.
2. https://unesdoc.unesco.org/ark:/48223/pf0000372991
3. The history of digital facial coding is one inextricable from ad-testing and media research to try to understand how elements and narrative will be received by viewers. Disney have also teamed with researchers from Caltech and Simon Fraser University to gauge reactivity of facial expressions of movie-watching audiences (Deng et al. 2017).

CHAPTER 5

1. Data found through Carbon Intensity API at https://carbonintensity.org.uk/.
2. Available at https://mlco2.github.io/impact/.

3. See: https://quantum.ieee.org/images/files/pdf/ieee-framework-for-metrics-and-benchmarks-of-quantum-computing.pdf.

4. The idea for this is in reference to Article 5(1)(c) of the GDPR and Article 4(1)(c) of the EU's GDPR where personal data must be 'adequate, relevant and limited to what is necessary in relation to the purposes for which they are processed'.

5. Japanese names are historically written with the family name preceding their given name. Thus, the author Nishida Kitarō will be cited by the family name Nishida. Where an author publishes family name last, they will be cited and recorded in the reference list accordingly.

Chapter 6

1. See: https://www.4littletrees.com/.

2. See: https://www.dreambox.com/teacher/solutions/adaptive-learning.

3. See https://www.class.com/higher-ed/.

4. Available at https://www.ohchr.org/Documents/HRBodies/CRC/GCChildrensDigitalEnvironment/2020/others/emotional-ai-lab-2020-11-11.docx.

Chapter 7

1. Video available from Advertising Age at: https://adage.com/creativity/work/volvo-terrible-idea/2287851.

2. This helps the various elements and systems in a car communicate, particularly where one subsystem of the overall system may need to control actuators or receive feedback from sensors.

3. See https://www.automotiveworld.com/news-releases/affectiva-and-nuance-to-bring-emotional-intelligence-to-ai-powered-automotive-assistants/.

4. Dates and links to seminars are: 12/11/20 https://go.affectiva.com/road-safety-webinar; 10/12/20 https://go.affectiva.com/advanced-road-safety-webinar; 11/02/21 https://go.affectiva.com/automotive-occupant-experience-webinar.

Chapter 8

1. See: https://www.pymetrics.ai/assessments#core-games.

2. See: https://aws.amazon.com/partners/amazon-connect-and-salesforce/.

3. See: https://blog.zoom.us/zoom-iq-for-sales/.

4. See: https://www.uniphore.com/emotional-intelligence-products/.

5. Finance data found at Crunchbase. See: https://www.crunchbase.com/organization/uniphore.

6. See: https://cogitocorp.com/product/#.

7. See: https://www.vocitec.com/.

8. See: https://twitter.com/WolfieChristl.

9. A point recognised by Europe's data protection instruments. See for example Article 29's Opinion 2/2017 on data processing at work.

10. See: https://blog.affectiva.com/the-emotion-behind-your-online-conversation-transforming-the-qual-experience.
11. See: https://emojipedia.org/people/.

CHAPTER 9

1. See video from Kamitani's Lab of the presented image and the reconstructed images, decoded from brain activity. https://www.youtube.com/watch?v=jsp1KaM-avU.
2. See 48mins in https://www.youtube.com/watch?v=DVvmgjBL74w.
3. See https://choosemuse.com/legal/.
4. See https://readabilityformulas.com/free-readability-formula-tests.php.
5. See https://www.statista.com/statistics/264810/number-of-monthly-active-facebook-users-worldwide/.

CHAPTER 10

1. Except in the case of mental health charities, those not working, the retired and house persons; and except in the case of academic research, the retired and house persons.

Aas, Sean, and David Wasserman. 2016. "Brain–computer interfaces and disability: Extending embodiment, reducing stigma?" *Journal of Medical Ethics*, 42 no.1: 37–40.

Achterhuis, Hans. 2001. "Andrew Feenberg: Farewell to Dystopia." In *American Philosophy of Technology: The Empirical Turn*, edited by Hans Achterhuis, 65–93. Bloomington: Indiana University Press.

Adler-Bell, Sam, and Michelle Miller. 2018. "The Datafication of Employment." Last modified April 22, 2022. https://tcf.org/content/report/datafication-employment-surveillance-capitalism-shaping-workers-futures-without-knowledge/.

Adorno, Theodor W. 1976 [1969]. "Introduction." In *The Positivist Dispute in German Sociology*, edited by Theodor W. Adorno, Hans Albert, and Ralf Dahrendorf, 1–67. London: Heinemann Educational Books Ltd.

Ahmed, Sara. 2010. *The Promise of Happiness*. Durham: Duke University Press.

Ahmed, Sara. 2014. *The Cultural Politics of Emotion*. Edinburgh: Edinburgh University Press.

AI Index. 2019. "2019 Report." Last modified April 22, 2022. https://hai.stanford.edu/sites/default/files/ai_index_2019_report.pdf

AI Now Institute. 2019. "2019 Report." Last modified April 22, 2022. https://ainowinstitute.org/AI_Now_2019_Report.pdf

Amazon AWS. 2023. "Emotion." Last modified July 5, 2023. https://docs.aws.amazon.com/rekognition/latest/APIReference/API_Emotion.html

Anderson, Elizabeth. 2020. "Feminist Epistemology and Philosophy of Science, Stanford Encyclopaedia of Philosophy." Last modified April 22, 2022. https://plato.stanford.edu/entries/feminism-epistemology/

Andrejevic, Mark. 2013. *Infoglut: How Too Much Information is Changing the Way We Think and Know*. New York: Routledge.

Andrejevic, Mark. 2020. *Automated Media*. New York: Routledge.

Andrejevic, Mark, and Neil Selwyn. 2020. "Facial recognition technology in schools: critical questions and concerns." *Learning, Media and Technology*, 45 no.2: 115–128. https://doi.org/10.1080/17439884.2020.1686014

Agüera y Arcas, Blaise, Mitchell, Margaret, and Alexander Todorov. 2018. "Physiognomy's New Clothes Medium." Last modified April 22, 2022. https://medium.com/@blaisea/physiognomys-new-clothes-f2d4b59fdd6a

Arnold, Madga. B. 1960. "Emotion and Personality." *Psychological Aspects*, *Vol. 1*. New York: Columbia University Press.

Arthur, Joeanna. (n.d.). "Neural Engineering System Design (NESD)." Last modified July 4, 2023. https://www.darpa.mil/program/neural-engineering-system-design

Article 19. 2021. "Emotional Entanglement: China's emotion recognition market and its implications for human rights." Last modified April 22, 2022. https://www.article19.org/wp-content/uploads/2021/01/ER-Tech-China-Report.pdf

Automotive World. 2021. "Bosch at the IAA Mobility: Safe, emissions-free, and exciting mobility – now and in the future." Last modified April 22, 2022. https://www.automotiveworld.com/news-releases/bosch-at-the-iaa-mobility-safe-emissions-free-and-exciting-mobility-now-and-in-the-future/

Ayoub, Jackie, Zhou, Feng, Bao, Shan, and X. Jessie Yang. 2019. "From Manual Driving to Automated Driving: A Review of 10 Years of AutoUI." In Proceedings of the 11th International Conference on Automotive User Interfaces and Interactive Vehicular Applications (AutomotiveUI '19). Association for Computing Machinery, 70–90. New York: USA. https://doi.org/10.1145/3342197.3344529

Azari, Bahar, Westlin, Christiana, Satpute, Ajay B, Hutchinson, J. Benjamin, Kragel, Philip A, Hoemann, Katie, Khan, Zulqarnain, Wormwood, Jolie B, Quigley, Karen S, Erdogmus, Deniz, Dy, Jennifer, Brooks, Dana H, and Lisa Feldman Barrett. 2020. "Comparing supervised and unsupervised approaches to emotion categorization in the human brain, body, and subjective experience." *Scientific Reports*, 10 no. 20284: 1–17. https://doi.org/10.1038/s41598-020-77117-8

Bailenson, Jeremy N. 2021. "Nonverbal overload: A theoretical argument for the causes of Zoom fatigue." *Technology, Mind, and Behavior*, 2 no. 1. https://doi.org/10.1037/tmb0000030

Bakir, Vian. 2013. *Torture, Intelligence and Sousveillance in the War on Terror: Agenda-Building Struggles*. Farnham: Ashgate.

Bakir, Vian, and Andrew McStay. 2022. *Optimising Emotions, Incubating Falsehoods: How to Protect the Global Civic Body from Disinformation and Misinformation*. Basingstoke: Palgrave.

Bakir, Vian. and Andrew McStay. 2016. "Theorising Transparency Arrangements: Assessing Interdisciplinary Academic and Multi-Stakeholder Positions on Transparency in the post-Snowden Leak Era." *Ethical Space: Journal of Communication*, 3 no. 1: 24–31.

Bakir, Vian, McStay, Andrew, and Alex Laffer. 2021a. *UK Attitudes Towards Personal Data Stores and Control Over Personal Data, 2021*. [Data Collection]. Colchester, Essex: UK Data Service. 10.5255/UKDA-SN-855178

Bakir, Vian, McStay, Andrew, and Alex Laffer. 2021b. "Final Report Taking Back Control of Our Personal Data: An ethical impact assessment of personal data storage apps." Last modified April 22, 2022. https://drive.google.com/file/d/1yA5CifVAAOM2DNiS4MbBTmQ_r3FwSYji/view?usp=sharing

Baldwin, Robert C. 1940 "The Meeting of Extremes in Recent Esthetics." *The Journal of Philosophy*, 37 no. 13: 348–358.

Ball, Matthew. 2020. "The Metaverse: What It Is, Where to Find it, Who Will Build It, and Fortnite." Last modified April 22, 2022. https://www.matthewball.vc/all/theme taverse

Banbury, Colby R, Reddi, Vijay Janapa, Lam, Max, Fu, William, Fazel, Amin, Holleman, Jeremy, Huang, Xinyuan, Hurtado, Robert, Kanter, David, Lokhmotov, Anton, Patterson, David, Pau, Danilo, Seo, Jae-sun, Sieracki, Jeff, Thakker, Urmish, Verhelst, Marian, and Poonam Yadav. 2020. "Benchmarking TinyML systems: Challenges and direction." *arXiv*, 1–8. https://doi.org/10.48550/arXiv.2003.04821

Barrett Lisa Feldman. 2006 "Are Emotions Natural Kinds?" *Perspectives on Psychological Science*. 1 no.1: 28–58. https://doi.org/10.1111/j.1745-6916.2006.00003.x

Barrett, Lisa Feldman. 2017. *How Emotions are Made: The Secret Life of the Brain*. Boston: Houghton Mifflin Harcourt.

Barrett, Lisa Feldman, Adolphs, Ralph, Marsella, Stacy, Martinez, Aleix M, and Seth D. Pollak (2019) "Emotional Expressions Reconsidered: Challenges to Inferring Emotion From Human Facial Movements." *Psychological Science in the Public Interest*, 20 no.1: 1–68. https://doi.org/10.1177/1529100619832930

Barry, Laurence, and Arthur Charpentier 2020. "Personalization as a promise: Can Big Data change the practice of insurance?" *Big Data & Society*, 7 no.1. https://doi.org/10.1177/2053951720935143

Baudrillard, Jean. 1993 [1976]. *Symbolic Exchange and Death*. London: Sage.

Beauchamp, Tom L, and James F. Childress. 2001 [1979]. *Principles of Biomedical Ethics*. New York: Oxford University Press

Beauvisage, Thomas, and Kevin Mellet. 2020. "Datassets: Assetizing and Marketizing Personal Data." In *Turning Things into Assets*, edited by Kean Birch and Fabian Muniesa, 75–95. Cambridge, MA: MIT Press.

Bender, E.M.; Gebru, T.; McMillan-Major, A.S. and Shmitchell, S. (2021) On the Dangers of Stochastic Parrots: Can Language Models Be Too Big? In *Conference on Fairness, Accountability, and Transparency (FAccT '21)*, March 3–10, 2021, Virtual Event, Canada. ACM, 610–623. New York, NY, USA.

Benjamin, Ruha. 2019. *Race After Technology*. Cambridge: Polity.

Bennett, Maxwell, Dennett, Daniel, Hacker, Peter, John Searle. 2007. *Neuroscience and Philosophy: Brain, Mind and Language*. New York: Columbia University Press.

Bentham, Jeremy. 1834. *Deontology: or, the Science of Morality*. London: Longman, Rees, Orme, Brown, Green & Longman.

Berardi, Franco. 2009. *The Soul at Work: From Alienation to Autonomy*. Los Angeles, CA: Semiotext(e).

Berg, Martin. 2022. "HATE IT? AUTOMATE IT! Thinking and doing robotic process automation and beyond." In *Everyday Automation Experiencing and Anticipating Emerging Technologies*, edited by Sarah Pink, Martin Berg, Deborah Lupton and Minna Ruckenstein, 157–170. London: Routledge.

Birhane, Abeba. 2021a. "Algorithmic injustice: a relational ethics approach." *Patterns*, 2 no.2: 1–9. https://doi.org/10.1016/j.patter.2021.100205.

Birhane, Abeba. 2021b. "Cheap AI." In *Fake AI*, edited by Frederike Kaltheuner, 41–51. Meatspace.

Birhane, Abeba. 2021c. "The Impossibility of Automating Ambiguity." *Artif Life*, 27 no.1: 44–61. https://doi.org/10.1162/artl_a_00336

Boehner, Kirsten, Sengers, Phoebe, and Simeon Warner. 2008. "Interfaces with the Ineffable: Meeting Aesthetic Experience on its Own Terms." *ACM Transactions on Computer-Human Interaction* 12: 1–29. https://doi.org/10.1145/1453152.1453155

Boenick, Marianne, Swierstra, Tsjalling, and Dirk Stemerding. 2010. "Anticipating the Interaction between Technology and Morality: A Scenario Study of Experimenting with Humans in Bionanotechnology." *Studies in Ethics, Law, and Technology* 4 no.2, Article 4. https://doi.org/10.2202/1941-6008.1098

Bogen, Miranda, and Aaron Rieke. 2018. "Help Wanted: An Examination of Hiring Algorithms, Equity, and Bias." *Upturn*, 1–75. Last modified April 22, 2022. https://www.upturn.org/static/reports/2018/hiring-algorithms/files/Upturn%20--%20Help%20Wanted%20-%20An%20Exploration%20of%20Hiring%20Algorithms,%20Equity%20and%20Bias.pdf

Bonilla-Silva, Eduardo. 2019. "Feeling Race: Theorizing the Racial Economy of Emotions." *American Sociological Review*, 84 no.1: 1–25. https://doi.org/10.1177/0003122418816958

Bosch. 2022a. "Interior Monitoring Systems." Last modified April 22, 2022. https://www.bosch-mobility-solutions.com/en/solutions/interior/interior-monitoring-systems/

Bosch. 2022b. "Connected Mobility." Last modified April 22, 2022. https://www.bosch-mobility-solutions.com/en/mobility-topics/connected-mobility/

Bösel, Bernd, and Serjoscha Wiemer, eds. 2020. *Affective Transformations: Politics, Algorithms, Media.* Lüneburg: Meson Press.

Bostrom, Nick, and Toby Ord. 2006. "The Reversal Test: Eliminating Status Quo Bias in Applied Ethics." *Ethics.* 116: 656–679. https://doi: 10.1086/505233.

Brey, Philip A.E. 2008. "The Technological Construction of Social Power." *Social Epistemology: A Journal of Knowledge, Culture and Policy*, 22 no.1: 71–95. https://doi.org/10.1080/02691720701773551

Brey, Philip A.E. 2012. "Anticipating ethical issues in emerging IT." *Ethics and Information Technology*, 14: 305–317. https://doi.org/10.1007/s10676-012-9293-y

Broad, Charlie D. (1954) "Emotion and Sentiment." *Journal of Aesthetics and Art Criticism*, 13 no.2: 203–214. https://doi.org/10.1111/1540_6245.jaac13.2.0203

Browne, Simone. 2015. *Dark matters: on the surveillance of blackness.* Durham: Duke University Press.

Brunton, Finn, and Helen Nissenbaum. 2015. *Obfuscation: A User's Guide for Privacy and Protest.* Cambridge, MA: MIT Press.

Bublitz, Jan Christoph, Reinhard Merkel. 2014. "Crime Against Minds: On Mental Manipulations, Harms and a Human Right to Mental Self-Determination." *Criminal Law and Philosophy.* 8: 51–77. https://doi.org/10.1007/s11572-012-9172-y

Buolamwini, Joy, and Gebru, Timnit. 2018. "Gender Shades: Intersectional Accuracy Disparities in Commercial Gender Classification." *Proceedings of the 1st Conference on Fairness, Accountability and Transparency (Proceedings of Machine Learning Research)*, Sorelle A. Friedler and Christo Wilson (Eds.), PMLR 81: 77–91. Last modified April 22, 2022. http://proceedings.mlr.press/v81/buolamwini18a.html.

Butch, Richard. 2000. *The Making of American Audiences: From Stage to Television, 1750–1990*. Cambridge: Cambridge University Press.

Butler, Sarah 2021. "Gig-working in England and Wales more than doubles in five years." *The Guardian*. Last modified May 30, 2023. https://www.theguardian.com/business/2021/nov/05/gig-working-in-england-and-wales-more-than-doubles-in-five-years.

Cabinet Office. 2019. "Social Principles of Human-centric AI (Draft)." Last modified April 22, 2022. https://www8.cao.go.jp/cstp/stmain/aisocialprinciples.pdf

Cabitza, Federico, Campagner, Andrea, and Martina Mattioli. 2022. "The unbearable (technical) unreliability of automated facial emotion recognition." *Big Data & Society*, 9(2). https://doi.org/10.1177/20539517221129549

Carter, Robert E. 2013. *The Kyoto School: An Introduction*. Albany: SUNY.

Caswell, E. 2015. "Color film was built for white people. Here's what it did to dark skin." *Vox*. Last modified April 22, 2022. https://www.vox.com/2015/9/18/9348821/photography-race-bias

CDEI. 2021a. "AI Barometer Part 5 – Education." Last modified April 22, 2022. https://www.gov.uk/government/publications/ai-barometer-2021/ai-barometer-part-5-education

CDEI. 2021b. Unlocking the value of data: Exploring the role of data intermediaries. Last modified April 22, 2022. https://www.gov.uk/government/publications/unlocking-the-value-of-data-exploring-the-role-of-data-intermediaries/unlocking-the-value-of-data-exploring-the-role-of-data-intermediaries

Chakravorti, Bhaskar, Bhalla, Ajay, and Ravi Shankar Chaturvedi. 2021. "How Digital Trust Varies Around the World." *Harvard Business Review*. Last modified April 22, 2022. https://hbr.org/2021/02/how-digital-trust-varies-around-the-world

Charitsis, Vassilis, Zwick, Detlev, Alan Bradshaw. 2018. "Creating worlds that create audiences: Theorising personal data markets in the age of communicative capitalism." *tripleC: Communication, Capitalism & Critique*. 16 no.2: 820–34. https://doi.org/10.31269/triplec.v16i2.1041

Chekroud, Adam M, Everett, Jim A. C, Bridge, Holly, Miles Hewstone (2014) "A review of neuroimaging studies of race-related prejudice: does amygdala response reflect threat?" Frontiers in Human Neuroscience, 8: 179. https://doi.org/10.3389/fnhum.2014.00179

Chen, Ted Mo. 2021. "Edtech will survive China's crackdown, but it won't be the same." *Technode*. Last modified April 22, 2022. https://technode.com/2021/08/31/EdTech-survive-chinas-crackdown-but-it-wont-be-the-same/

Chinoy, Sahil. 2019. "The Racist History Behind Facial Recognition." *The New York Times*. Last modified April 22, 2022. https://www.nytimes.com/2019/07/10/opinion/facial-recognition-race.html

Christl, Wolfie. 2021. "Digital surveillance and control in the workplace: From expanding operational data collection to algorithmic management? A study by Cracked Labs." Last modified April 22, 2022. https://crackedlabs.org/daten-arbeitsplatz

Clarke, Adele E, Mamo, Laura, Fosket, Jennifer Ruth, Fishman, Jennifer R, Shim, and Janet K. 2010. *Biomedicalization: Technoscience, Health and Illness in the U.S.* Durham and London: Duke University Press.

Clauss, Ludwig Ferdinand. 1929. *Von Seele und Antlitz der Rassen und Völker: eine Einführung in die vergleichende Ausdrucksforschung.* München: Lehmann.

Clynes, Manfred. 1977. *Sentics: The Touch of the Emotions.* New York: Anchor Press/Doubleday.

Coeckelbergh, Mark. 2020. *AI Ethics.* Cambridge, Massachusetts: MIT Press.

Cohen, Julie E. 2019. *Between Truth and Power: The Legal Constructions of Informational Capitalism.* Oxford: Oxford University Press.

Cooper, Bridget. 2011. *Empathy in Education: Engagement, Values and Achievement.* London: Continuum.

Costanza-Chock, Sasha. 2020. *Design Justice: Community-Led Practices to Build the Worlds We Need.* Cambridge, MA: MIT.

Council of Europe. 2021. "CONSULTATIVE COMMITTEE OF THE CONVENTION FOR THE PROTECTION OF INDIVIDUALS WITH REGARD TO AUTOMATIC PROCESSING OF PERSONAL DATA." Last modified April 22, 2022. https://rm.coe.int/guidelines-on-facial-recognition/1680a134f3

Cowen, Alan, Sauter, Disa, Tracy, Jessica L, Dacher Keltner 2019. "Mapping the Passions: Toward a High-Dimensional Taxonomy of Emotional Experience and Expression." *Psychological Science in the Public Interest* 20 no.1: 69–90.

Crampton, Natasha. 2022. *Microsoft's framework for building AI systems responsibly.* Last modified January 31, 2023. https://blogs.microsoft.com/on-the-issues/2022/06/21/microsofts-framework-for-building-ai-systems-responsibly/

Crawford, Kate. 2021. *The Atlas of AI: Power, Politics, and the Planetary Costs of Artificial Intelligence.* New Haven, CT: Yale University Press.

Crawford, Kate. and Vladan Joler. 2018. "Anatomy of an AI system." *Virtual Creativity* 9 no.1. 117–120. https://doi.org/10.1386/vcr_00008_7

Critchley, Simon. 1999. *Ethics, Politics, Subjectivity.* London: Verso.

Crutzen, Paul. 2002. "Geology of mankind." *Nature* 415 no. 6867: 23. https://doi.org/10.1038/415023a.

Cunningham, William A, Johnson, Marcia K, Raye, Carol L, Gatenby, J. Chris, Gore, John C, and Mahzarin R. Banaji, 2004. "Separable neural components in the processing of black and white faces." *Psychological Science.* 15, 806–813. doi: 10.1111/j.0956-7976.2004.00760.x

Damasio, Antonio R. 1999. *The feeling of what happens: Body and emotion in the making of consciousness.* New York: Harcourt Brace.

Damasio, Antonio R. 2003. *Looking for Spinoza: Joy, Sorrow, and the Feeling Brain.* Orlando: Harcourt.

Daniela, Linda. 2020. *New Perspectives on Virtual and Augmented Reality.* London: Taylor & Francis.

Darrow, Chester W. 1934. "The Reflexohmeter (Pocket Type)." *The Journal of General Psychology* 10 no.238. https://doi.org/10.1080/00221309.1934.9917731

Darwin, Charles. 2009 [1872]. *The Expression of the Emotions in Man and Animals.* London: Harper.

Data Dividend Project. 2021. "Homepage." Last modified April 22, 2022. https://www.datadividendproject.com/

Daub, Adrian. 2018. "The Return of the Face." *Longreads*. Last modified April 22, 2022. https://longreads.com/2018/10/03/the-return-of-the-face/

Davis, Fred D. 1985. "A technology acceptance model for empirically testing new end-user information systems: Theory and results." Last modified April 22, 2022. https://dspace.mit.edu/handle/1721.1/15192

Davies, William. 2015. *The Happiness Industry: How the Government & Big Business Sold us Wellbeing*. London: Verso.

DCMS. 2019. *UK National Data Strategy*. Last modified April 22, 2022. https://www.gov.uk/government/publications/uk-national-data-strategy.

De Bot, Dirk. Tom Haegemans. 2021. "Data Sharing Patterns as a Tool to Tackle Legal Considerations about Data Reuse with Solid: Theory and Applications in Europe." Last modified April 22, 2022. https://lirias.kuleuven.be/retrieve/599839

DeCew, Judith Wagner. 1997. *In Pursuit of Privacy: Law, Ethics, and the Rise of Technology*. New York: Cornell University Press.

De Certeau, Michel. 1988. *The Practice of Everyday Life* Berkeley, CA: University of California Press.

Defence and Security Accelerator. 2022. "Open Call Innovation Focus Areas." *Gov. UK*. Last modified April 22, 2022. https://www.gov.uk/government/publications/defence-and-security-accelerator-dasa-open-call-for-innovation/open-call-innovation-focus-areas

Dembour, Marie-Bénédicte. 2010 "What Are Human Rights? Four Schools of Thought." *Human Rights Quarterly* 32 no.1: 1–20. https://doi.org/10.1353/hrq.0.0130

Dencik, Lina. 2021. "Towards data justice unionism? A labour perspective on AI governance." In *AI for Everyone? Critical Perspectives*, edited by. Pieter Verdegem, 267–84. London: University of Westminster Press.

Dencik, Lina and Cable, Jonathan. 2017. "The Advent of Surveillance Realism: Public Opinion and Activist Responses to the Snowden Leaks." *International Journal of Communication* 11: 763–781.

Deng, Zhiwei, Navarathna, Rajitha, Carr, Peter, Mandt, Stephan, Yue, Yisong, Matthews, Iain, and Greg Mori. (2017) "Factorized Variational Autoencoders for Modeling Audience Reactions to Movies." 2017 IEEE Conference on Computer Vision and Pattern Recognition (CVPR), 6014–6023. http://doi.org/10.1109/CVPR.2017.637.

Denton, Emily, Hanna, Alex, Amironesei, Razvan, Smart, Andrew, Hilary Nicole 2021. "On the genealogy of machine learning datasets: A critical history of ImageNet." *Big Data & Society*. https://doi.org/10.1177/20539517211035955

Dewey, John. 1894. "The theory of emotion: I: Emotional attitudes." *Psychological Review* 1 no.6: 553–569. https://doi.org/10.1037/h0069054

Digital Defend Me. 2022. "The State of Biometrics 2022: A Review of Policy and Practice in UK Education." Last modified April 22, 2022. https://defenddigitalme.org/research/state-biometrics-2022/#foreword

Digi.me. 2022. "Ubdi." Last modified April 22, 2022. https://digi.me/ubdi/#slide-0

Digital Futures Commission. 2021. "Governance of data for children's learning in UK state schools." Last modified April 22, 2022. https://digitalfuturescommission.org.uk/wp-content/uploads/2021/06/Governance-of-data-for-children-learning-Final.pdf

Dignum, Virginia. 2022 *Relational Artificial Intelligence*, 1–15. https://arxiv.org/pdf/2202.07446.pdf

Dixon, Thomas. 2012. "'Emotion': The History of a Keyword in Crisis." *Emotion Review*, 4 no.4: 338–344. https://doi.org/10.1177/1754073912445814

D'Mello, Sidney K. 2017 "Emotional learning analytics." In *Handbook of Learning Analytics & Educational Data Mining*, edited by Charles Lang, George Siemens, Alyssa Wise, and Dradon Gašević, 115–27. Edmonton, AB: Society for Learning Analytics Research.

D'Mello, Sidney, Kappas, Arvid, Jonathan Gratch 2018. "The Affective Computing Approach to Affect Measurement." *Emotion Review* 10 no.2: 174–183. https://doi.org/10.1177/1754073917696583

D'Mello, Sidney, Art Graesser. 2012. "AutoTutor and Affective AutoTutor: Learning by talking with cognitively and emotionally intelligent computers that talk back." *ACM Transactions on Interactive Intelligent Systems* 2 no.4: 23:22–23:39. https://doi.org/10.1145/2395123.2395128

Di Paolo, Ezequiel A. 2000. "Homeostatic adaptation to inversion of the visual field and other sensorimotor disruptions." Last modified April 22, 2022. http://users.sussex.ac.uk/~inmanh/adsys10/Seminars/Seminar_Ashby/homeo.pdf

Di Paolo, Ezequiel A. 2003. "Organismically-Inspired Robotics: Homeostatic Adaptation and Teleology Beyond the Closed Sensorimotor Loop." Last modified April 22, 2022. http://users.sussex.ac.uk/~ezequiel/dp-erasmus.pdf

Dorrestijn, Steven. 2012. "Technical Mediation and Subjectivation: Tracing and Extending Foucault's Philosophy of Technology." *Philosophy & Technology* 25: 221–241. https://doi.org/10.1007/s13347-011-0057-0

Dourish, Paul. 2019. "User experience as legitimacy trap." *Interactions* 26 no.6: 46–49. https://doi.org/10.1145/3358908

Draper, Nora A. 2019. *The Identity Trade: Selling Privacy and Reputation Online*. New York: New York University Press.

Dreyfus, Hubert L. 1991. *Being-in-the-World: A Commentary on Heidegger's Being and Time, Division 1*. Cambridge, MA: MIT.

Dreyfus, Hubert L, and Stuart E. Dreyfus. 1990. "Making a Mind Versus Modelling the Brain: Artificial Intelligence Back at a Branch-point." In *The Philosophy of Artificial Intelligence*, edited by Margaret Boden, 309–333. Oxford: OUP.

Dror, Otniel E. 1999. "The Scientific Image of Emotion: Experience and Technologies of Inscription." *Configurations* 7: 355–401. https://doi.org/10.1353/con.1999.0025

Dror, Otniel E. 2001. Counting the Affects: Discoursing in Numbers, *Social Research*, 68 no. 2: 357–378. https://doi.org/10.1023/a:1017336822217

Duchenne de Boulogne, Guillaume-Benjamin-Amand. 1990 [1862]. *The Mechanism of Human Facial Expression*. Cambridge: Cambridge University Press.

Dzieza, Josh. 2020. "HOW HARD WILL THE ROBOTS MAKE US WORK?" *The Verge*. Last modified April 22, 2022. https://www.theverge.com/2020/2/27/21155254/automation-robots-unemployment-jobs-vs-human-google-amazon

EDPB. 2020. "Guidelines 1/2020 on processing personal data in the context of connected vehicles and mobility related applications." Last modified April 22, 2022. https://edpb.

europa.eu/sites/default/files/consultation/edpb_guidelines_202001_connectedvehic
les.pdf

EDPS. 2020. "Data Protection." Last modified April 22, 2022. https://edps.europa.eu/
data-protection_en

Ekman, Paul. 1977. "Biological and cultural contributions to body and facial
movement." In *The anthropology of the body*, edited by John Blacking, 39–84.
London: Academic Press,

Ekman, Paul. 1989. "The Argument and Evidence About Universals in Facial Expressions
of Emotions." In *Handbook of Social Psychophysiology*, edited by Hugh Wagner and
Anthony Manstead, 143–64. Chichester: Wiley.

Ekman, Paul. 1992. "An argument for basic emotions." *Cognition and Emotion*, 6 no.3–
4: 169–200. https://doi.org/10.1080/02699939208411068

Ekman, Paul, Wallace V. Friesen. 1969. "The Repertoire of Nonverbal Behavior: Categories,
Origins, Usage, and Coding." *Semiotica*, 1 no.1: 49–98. https://doi.org/10.1515/
semi.1969.1.1.49

Ekman, Paul, Wallace V. Friesen. 1971. "Constants across cultures in the face and emotion." *Journal of Personality and Social Psychology* 17 no.2: 124–129. https://doi.org/
10.1037/h0030377

Ekman, Paul, Wallace V. Friesen. 1978. *Facial Action Coding System: A Technique for
the Measurement of Facial Movement*. Palo Alto, CA: Consulting Psychologists Press.

Elias, Maurice J, Zins, Joseph E, Weissberg, Roger P, Frey, Karin S, Greenberg, Mark T,
Haynes, Norris M, Kessler, Rachael, Schwab-Stone, Mary E, and Timothy P. Shriver.
1997. *Promoting Social and Emotional Learning: Guidelines for Educators*. Alexandria,
Virginia: Association for Supervision and Curriculum Development. Last modified
April 22, 2022. https://earlylearningfocus.org/wp-content/uploads/2019/12/promot
ing-social-and-emotional-learning-1.pdf

Element AI & Nesta. 2019. "Data Trusts A new tool for data governance." Last modified
April 22, 2022. https://hello.elementai.com/rs/024-OAQ-547/images/Data_Trusts_
EN_201914.pdf

Emejulu, Akwugo, and Francesca Sobande. 2019. "Afrofeminism and Black Feminism
in Europe." Last modified April 22, 2022. https://www.plutobooks.com/blog/europe-
black-feminism-afrofeminism-emejulu/

Ess, Charles M. 2020. "Interpretative Pros Hen Pluralism: from Computer-Mediated
Colonization to a Pluralistic Intercultural Digital Ethics." *Philosophy & Technology*
33: 551–569. https://doi.org/10.1007/s13347-020-00412-9

European Commission. 2020a. "EU ROAD SAFETY POLICY FRAMEWORK 2021 -
2030 Next steps towards 'Vision Zero'." Last modified April 22, 2022. https://op.eur
opa.eu/en/publication-detail/-/publication/d7ee4b58-4bc5-11ea-8aa5-01aa75ed71a1

European Commission. 2020b. "Proposal for a REGULATION OF THE EUROPEAN
PARLIAMENT AND OF THE COUNCIL on European data governance (Data
Governance Act)." Last modified April 22, 2022. https://eur-lex.europa.eu/legal-cont
ent/EN/TXT/?uri=CELEX%3A52020PC0767

European Commission. 2021. "Proposal for a Regulation Laying Down Harmonised
Rules on Artificial Intelligence (Artificial Intelligence Act)." Last modified April 22,

2022. https://digital-strategy.ec.europa.eu/en/library/proposal-regulation-laying-down-harmonised-rules-artificial-intelligence

European Commission. 2023. "DRAFT Compromise Amendments on the Draft Report: Proposal for a regulation of the European Parliament and of the Council on harmonised rules on Artificial Intelligence (Artificial Intelligence Act) and amending certain Union Legislative Acts." Last modified July 1, 2023. https://www.europarl.eur opa.eu/meetdocs/2014_2019/plmrep/COMMITTEES/CJ40/DV/2023/05-11/Consol idatedCA_IMCOLIBE_AI_ACT_EN.pdf

Euro NCAP. 2018. "Euro NCAP 2025 Roadmap." Last modified April 22, 2022. https://cdn.euroncap.com/media/30700/euroncap-roadmap-2025-v4.pdf

European Parliament. 2018. "Vehicle Safety Regulation 2018/0145." Last modified April 22, 2022. https://data.consilium.europa.eu/doc/document/PE-82-2019-INIT/en/pdf

Expert Group on Architecture for AI Principles to be Practiced. 2021. "AI Governance in Japan Ver. 1.0." *METI*, 1–33. Last modified April 22, 2022. https://www.meti.go.jp/press/2020/01/20210115003/20210115003-3.pdf

Faception. 2021. "Homepage." Last modified April 22, 2022. https://www.faception.com/

Finger, Stanley. 2001. *Origins of Neuroscience: A History of Explorations Into Brain Function*. New York: Oxford University Press.

Finn, Emily S, Shen, Xilin, Scheinost, Dustin, Rosenberg, Monica D, Huang, Jessica, Chun, Marvin M, Papademetris, Xenophon, R Todd Constable. 2015. "Functional connectome fingerprinting: identifying individuals using patterns of brain connectivity." *Nature Neuroscience* 18 no.11:1664–1671.

Floridi, Luciano. 2018. "Soft Ethics: Its Application to the General Data Protection Regulation and Its Dual Advantage." *Philosophy & Technology* 31: 163–167. https://doi.org/10.1007/s13347-018-0315-5

Forbes-Riley, Kate, Diane Litman. 2011. "Benefits and challenges of real-time uncertainty detection and adaptation in a spoken dialogue computer tutor." *Speech Communication* 53 no. 9–10: 1115–1136. https://doi.org/10.1016/j.specom.2011.02.006

Foot, Philippa. 1967. "The Problem of Abortion and the Doctrine of the Double Effect." *Oxford Review*. 5: 143–145. https://doi.org/10.2307/j.ctv19m64sz.11

Friedman, Milton. 1962 [2002]. *Capitalism and Freedom*. Chicago: Chicago University Press.

Fridlund, Alun J. 2017. "The behavioral ecology view of facial displays, 25 years later." In *The science of facial expression*, edited by James A. Russell, J. and José Miguel Fernandez-Dols, 77–92. Oxford Scholarship Online. https://doi.org/10.1093/acp rof:oso/9780190613501.003.0005

Fuchs, Christian. 2013. *Digital Labour and Karl Marx*. New York: Routledge.

Fuchs, C. 2021. The Digital Commons and the Digital Public Sphere: How to Advance Digital Democracy Today. Last modified April 22, 2022. https://www.westminsterpap ers.org/article/id/917/

Gal, Danit. 2020. "China's Approach to AI Ethics" In *The AI Powered State: China's Approach to Public Sector Innovation*, 1–76. Last modified April 22, 2022. https://media.nesta.org.uk/documents/Nesta_TheAIPoweredState_2020.pdf

Galton, Francis. 1885 *Composite Photographs of Jewish Faces*, Wellcome Collection. Last modified April 22, 2022. https://wellcomecollection.org/works/ngq29vyw/items?canvas=1

Galton, Francis. 1901. "Biometry." *Biometrika* 1 no.1: 7–10. https://doi.org/10.1093/biomet/1.1.7

Gangadharan, Seeta Pea, Jędrzej Niklas. 2019. "Decentering technology in discourse on discrimination." *Information, Communication & Society* 22 no.7: 882–899. https://doi.org/10.1080/1369118X.2019.1593484 [24]

Gazzola, Valeria, Aziz-Zadeh, Lisa, Christian Keysers (2006) "Empathy and the Somatotopic Auditory Mirror System in Humans." *Current Biology* 16 no.18: 1824–1829. https://doi.org/10.1016/j.cub.2006.07.072

George, Damian, Reutimann, Kento, Aurelia Tamò-Larrieux. 2019. "GDPR bypass by design? Transient processing of data under the GDPR." *International Data Privacy Law* 9 no.4: 285–298. https://doi.org/10.1093/idpl/ipz017

Ginzburg, Carlo 1989. *Clues, Myths, and the Historical Method*. Baltimore, MD: John Hopkins University Press.

Glannon, Walter. 2016. "Brain-computer interfaces in end-of-life decision-making." *Brain-Computer Interfaces*, 3 no.3: 133–139. https://doi.org/10.1080/2326263x.2016.1207496

Glenn, Andrea L, and Adrian Raine. 2014. "Neurocriminology: implications for the punishment, prediction and prevention of criminal behaviour." *Nature Reviews Neuroscience* 15 no. 1: 54–63.

Global Partnership on AI Report. 2021. "CLIMATE CHANGE AND AI: Recommendations for Government Action." Last modified April 22, 2022. https://www.gpai.ai/projects/climate-change-and-ai.pdf

Goldman, Alvin I. 2008. *Simulating Minds: The Philosophy, Psychology, and Neuroscience of Mindreading*. New York: Oxford University Press.

Gopinath, Sridhar,. Ghanathe, Nikhil, Seshadri, Vivek, and Rahul Sharma. 2019. "Compiling kb-sized machine learning models to tiny IOT devices." *Proceedings of the 40th ACM SIGPLAN Conference on Programming Language Design and Implementation*, 79–95. https://doi.org/10.1145/3314221.3314597

Gramelsberger, Gabriele. 2020. "Algorithm Awareness: Towards a Philosophy of Artifactuality." In *Affective Transformations: Poliitcs, Algorithms, Media*, edited by Bernd Bösel and Serjoscha Wiemer, 41–48. Meson Press.

Gray, Richard T. 2004. *About Face: German Physiognomic Thought from Lavater to Auschwitz*. Detroit: Wayne State University Press.

Greely, Henry T. 2009. "Neuroscience-Based Lie Detection: The Need for Regulation." In *Using Imaging to Identify Deceit: Scientific and Ethical Questions*, edited by Emilio Bizzi, Steven E. Hyman, Marcus E. Raichle, Nancy Kanwisher, Elizabeth Anya Phelps, Stephen J. Morse, Walter Sinnott-Armstrong, Jed S. Rakoff, and Henry T. Greely, 46–55. Cambridge, MA: American Academy of Arts and Sciences.

Gregory, Karenm, and Jathan Sadowski. 2021. "Biopolitical platforms: the perverse virtues of digital labour." *Journal of Cultural Economy* 14 no. 6: 662–674. https://doi.org/10.1080/17530350.2021.1901766

Grindrod, Peter. 2014. *Mathematical Underpinnings of Analytics: Theory and Applications*. Oxford: OUP Oxford.

Gross, Michael Joseph. 2018. "THE PENTAGON'S PUSH TO PROGRAM SOLDIERS' BRAINS" *The Atlantic*. Last modified April 22, 2022. https://www.theatlantic.com/magazine/archive/2018/11/the-pentagon-wants-to-weaponize-the-brain-what-could-go-wrong/570841/

Grossman, Elizabeth. 2007. *High Tech Trash: Digital Devices, Hidden Toxics, and Human Health*. Washington: Island Press.

Gunkel, David J. (2017) *The Machine Question: Critical Perspectives on AI, Robots, and Ethics*. Cambridge, Massachusetts: MIT.

Günther, Hans. 1927. *The Racial Elements of European History*. London: Metheun & Co.

Gupta, Abhishek. 2021. "The Imperative for Sustainable AI Systems." *The Gradient*. Last modified April 22, 2022. https://thegradient.pub/sustainable-ai/

Gurley, Lauren Kaori. (2021) *Amazon Delivery Drivers Forced to Sign 'Biometric Consent' Form or Lose Job*. Last modified April 22, 2022. https://www.vice.com/en/article/dy8n3j/amazon-delivery-drivers-forced-to-sign-biometric-consent-form-or-lose-job

Hagel III, John, and Marc Singer. 1999. *Net Worth: Shaping Markets When Customers Make the Rules*. Boston, MA: Harvard Business Review Press.

Hartman, Todd, Kennedy, Helen, Steedman, Robin, and Rhianne Jones. 2020. "Public perceptions of good data management: Findings from a UK-based survey." *Big Data & Society*, 7 no. 1: 1–16. https://doi.org/10.1177/2053951720935616

Haupt, Joachim. 2021. "Facebook futures: Mark Zuckerberg's discursive construction of a better world." New Media & Society 23 no.2: 237–257. https://doi.org/10.1177/1461444820929315

Hayek, Friedrich A. 2001 [1944]. *The Road to Serfdom*. Abingdon: Routledge.

Heidegger, Martin. 1993 [1954]. "The Question Concerning Technology," in M. Heidegger *Basic Writings*, ed. D.F. Krell, 311–341. New York: Harper Collins.

Heidegger, Martin. 2013 [1941–2]. *The Event*. Bloomington: Indiana University Press.

Heidegger, Martin. 2011 [1962]. *Being and Time*. New York: Harper & Row.

Hessami, Ali G, and Shaw, Patricia. (2021) "*Introductory Chapter: AI's Very Unlevel Playing Field*." In Factoring Ethics in Technology, Policy Making, Regulation and AI, edited by. Ali G, Hessami and Patricia Shaw, 1–9. IntechOpen.

High-Level Expert Group on Artificial Intelligence. 2018. "Draft Ethics Guidelines for Trustworthy AI." Last modified April 22, 2022. https://ec.europa.eu/newsroom/dae/document.cfm?doc_id=56433, 09/01/19.

Hildebrandt, Mireille. 2011. "Introduction: A multifocal view of human agency in the era of autonomic computing." In *Law, Human Agency, and Autonomic Computing: The Philosophy of Law Meets the Philosophy of Technology*, edited by Mireille Hildebrandt and Antoinette Rouvroy, 1–11. Abingdon: Routledge.

Hillman, Velislava. 2022. "Edtech procurement matters: It needs a coherent solution, clear governance and market standards." Last modified April 22, 2022. https://www.lse.ac.uk/social-policy/Assets/Documents/PDF/working-paper-series/02-22-Hillman.pdf

Hochschild, Arlie. 1983. *The Managed Heart: Commercialization of Human Feeling*. Berkeley: University of California Press.

Hongladarom, Soraj. 2016. "Intercultural information ethics: a pragmatic consideration." In *Information cultures in the digital age*, edited by Matthew Kelly and Jared
Bielby, 191–206. Wiesbaden: Springer.

Horikawa, Tomoyasu, and Yukiyasu Kamitani. 2017a. "Generic decoding of seen and
imagined objects using hierarchical visual features." *Nature Communications* 8
no.1: 1–15. https://doi.org/10.1038/ncomms15037.

Horikawa, Tomoyasu, and Yukiyasu Kamitani. 2017b. Hierarchical Neural
Representation of Dreamed Objects Revealed by Brain Decoding with Deep Neural
Network Features. *Frontiers in computational neuroscience* 11 no.4: 1–11. https://doi.
org/10.3389/fncom.2017.00004.

Huizinga, Johan. 1955 [1938]. *Homo Ludens*. Boston: Beacon.

Hume, David. 1896 [1739]. *A Treatise of Human Nature*. London: Oxford University Press.

Husserl, Edmund. 1982 [1913]. *Ideas Pertaining to a Pure Phenomenology and to
a Phenomenological Philosophy—First Book: General Introduction to a Pure
Phenomenology*. The Hague: Nijhoff.

Husserl, Edmund. 2002 [1952]. *Ideas Pertaining to a Pure Phenomenology and to a
Phenomenological Philosophy: Second Book*. Dordrecht: Kluwer.

IBM. 2005. "An architectural blueprint for autonomic computing." Last modified
April 22, 2022. https://www-03.ibm.com/autonomic/pdfs/AC%20Blueprint%20Wh
ite%20Paper%20V7.pdf

Ienca, Marcello, and Gianclaudio Malgieri. 2021. "Mental Data Protection and the
GDPR." *SSRC*. https://doi.org/10.2139/ssrn.3840403

Ihde, Don. 1990. *Technology and the Lifeworld: From Garden to Earth*. Bloomington,
IN: Indiana University Press.

Ihde, Don. 2002. *Bodies in Technology*. Minneapolis: Minnesota Press.

Ihde, Don. 2010. *Heidegger's Technologies: Postphenomenological Perspectives*.
New York: Fordham.

Intel Education. 2022. "Applying Artificial Intelligence to Transform How We Learn."
Last modified April 22, 2022. https://www.intel.co.uk/content/www/uk/en/educat
ion/transforming-education/ai-in-education.html?wapkw=emotion

International Rights Advocates. 2021. "Multinational companies are liable for human
rights abuses within their supply chains." Last modified April 22, 2022. https://www.
internationalrightsadvocates.org/cases/cobalt

Jaguar Land Rover. 2015. "JAGUAR LAND ROVER ROAD SAFETY RESEARCH
INCLUDES BRAIN WAVE MONITORING TO IMPROVE DRIVER
CONCENTRATION AND REDUCE ACCIDENTS." Last modified April 22, 2022.
https://www.jaguarlandrover.com/news/2015/06/jaguar-land-rover-road-safety-
research-includes-brain-wave-monitoring-improve-driver

James, William. (1884) What is an emotion? *Mind*, 9(43): 188–205.

Janssen, Heleen, Cobbe, Jennifer, and Jatinder Singh. 2020. "Personal information management systems: a user-centric privacy utopia?" *Internet Policy Review* 9 no.4: 1–25.
https://doi.org/10.14763/2020.4.1536

Jecker, Nancy S, Atiure. Caesar A, and Martin Odei Ajei. 2022. "The Moral Standing
of Social Robots: Untapped Insights from Africa." *Philosophy & Technology* 35 no.2.
https://doi.org/10.1007/s13347-022-00531-5

Kant, Immanuel. 1983 [1795]. *Perpetual Peace and Other Essays*. Indianapolis: Hackett.

Kassner, Rudolf. 1919. *Zahl and Gesicht: Nebst einer Einleitung: Der Umriß einer universalen Physiognomik*. Frankfurt: Suhrkamp.

Kaur, Harmanpreet, McDuff, Daniel, Williams, Alex C, Teevan, Jaime, Iqbal, Shamsi T. 2022. "I Didn't Know I Looked Angry": Characterizing Observed Emotion and Reported Affect at Work. In *CHI Conference on Human Factors in Computing Systems (CHI '22)*, April 29-May 5, 2022, New Orleans, LA, USA. ACM, New York, NY. https://doi.org/10.1145/3491102.3517453

Keyes, Os. 2018. "The Misgendering Machines: Trans/HCI Implications of Automatic Gender Recognition." *Proceedings of the ACM on Human-Computer Interaction Volume 2*, CSCW (Nov 2018), 2 no.8: 1–22. https://doi.org/10.1145/3274357

Kia. 2020. "Amplify Your Joy with Emotive Driving." Last modified April 22, 2022. https://www.kia.com/worldwide/future-technology.do

Koenecke, Allison, Nam, Andrew, Lake, Emily, Nudell, Joe, Quartey, Minnie, Mengesha, Zion, Toups, Connor, Rickford, John R, Jurafsky, Dan, and Sharad Goel. 2020. "Racial disparities in automated speech recognition." *Proceedings of the National Academy of Sciences* 117 no.14: 7684–7689. https://doi.org/10.1073/pnas.1915768117

Kornhuber, Hans H, and Lüder Deecke. 2016. "Brain potential changes in voluntary and passive movements in humans: readiness potential and reafferent potentials." *Pflügers Archiv: European Journal of Physiology* 468 no.7: 1115–1124. https://doi.org/10.1007/s00424-016-1852-3

Kret, Mariska E, Prochazkova, Eliska, Sterck, Elisabeth H.M, and Zanna Clay. 2020. "Emotional expressions in human and non-human great apes." *Neuroscience & Biobehavioral Reviews*, 115: 378 to 395

Lakoff, George, and Mark Johnson. 1999. *Philosophy in the Flesh: The Embodied Mind and Its Challenge to Western Thought*. New York: Basic Books.

Lanier, Jaron. 2013. *Who Owns the Future?* New York: Simon & Schuster.

Latour, Bruno. 1996. *Aramis, or the Love of Technology*. Cambridge, MA: Harvard University Press.

Latour, Bruno. 2004. *Politics of Nature: How to Bring the Sciences into Democracy*. Cambridge, MA: Harvard University Press.

Latour, Bruno. 2005. *Reassembling the Social: An Introduction to Actor-Network-Theory*. Oxford: Oxford University Press.

Latour, Bruno. 2013. *An Inquiry into Modes of Existence: An Anthropology of the Moderns*. Cambridge, MA: Harvard University Press.

Lavater, Johann C. 1792. *Physiognomy; or, the Corresponding Analogy between the Conformation of the Features and the Ruling Passions of the Mind*. London: Cowie, Lo, and Co.

Lavazza, Andrea. 2018. "Freedom of Thought and Mental Integrity: The Moral Requirements for Any Neural Prosthesis." *Frontiers in neuroscience* 12 no.82: 1–10. https://doi.org/10.3389/fnins.2018.00082

Law, John. 2012 [1987]. "Technology and Heterogeneous Engineering: The Case of Portuguese Expansion." In *The Social Construction of Technological Systems: New Directions in the Sociology and History of Technology*, edited by Wiebe E. Bijker, Thomas Parke Hughes, and Trevor Pinch, 105–127. Cambridge, MA: MIT.

Lawrence, Neil. 2016. "Data trusts could allay our privacy fears." *The Guardian*. Last modified April 22, 2022. https://www.theguardian.com/media-network/2016/jun/03/data-trusts-privacy-fears-feudalism-democracy

Lehtiniemi, Tuukka. 2017. "Personal Data Spaces: An Intervention in Surveillance Capitalism?" *Surveillance & Society* 15 no.5: 626–639. https://doi.org/10.24908/ss.v15i5.6424

Lehtiniemi, Tuukka, and Yki Kortesniemi. 2017. "Can the obstacles to privacy self-management be overcome? Exploring the consent intermediary approach." *Big Data & Society* 4 no.2. https://doi.org/10.1177/2053951717721935

Lemoine, Blake. 2022. "Is LaMDA Sentient?—an Interview." *Medium*. https://cajundiscordian.medium.com/is-lamda-sentient-an-interview-ea64d916d917

Lepper, Mark R, and Ruth W. Chabay. 1988. "Socializing the Intelligent Tutor: Bringing Empathy to Computer Tutors." In *Learning Issues for Intelligent Tutoring Systems*, edited by Heinz Mandl and Alan Lesgold, 242–257. New York: Springer-Verlag.

Leyffer, Sven, Wild, Stefan M, Fagan, Mike, Snir, Marc, Palem, Krishna, Yoshii, Kazutomo, and Hal Finkel. 2016. "Doing Moore with Less—Leapfrogging Moore's Law with Inexactness for Supercomputing." *Technical Report ANL/MCS-P6077-1016*. Argonne National Laboratory MCS. Last modified April 22, 2022. https://bibbase.org/network/publication/leyffer-wild-fagan-snir-palem-finkel-yoshii-doingmoorewithlessleapfroggingmooreslawwithinexactnessforsupercomputing-2018

Leys, Ruth. 2011. "The Turn to Affect: A Critique." *Critical Inquiry* 37 no.3: 434–472. https://doi.org/10.1086/659353

Liao, Tony. 2018. "Mobile versus headworn augmented reality: How visions of the future shape, contest, and stabilize an emerging technology." *New Media & Society* 20 no.2: 796–814. https://doi.org/10.1177/1461444816672019

Libet, Benjamin, Wright, Elwood W, and Curtis A Gleason. 1983. "Preparation- or intention-to-act, in relation to pre-event potentials recorded at the vertex Relation of the preparation or intention to perform an act with the potentials preceding the event collected at the vertex." *Electroencephalography and Clinical Neurophysiology* 56 no.4: 367–372. https://doi.org/10.1016/0013-4694(83)90262-6

Licklider, Joseph Carl Robnett. 1960. "Man-Computer Symbiosis." *IRE Transactions on Human Factors in Electronics*, HFE-1 no.1: 4–11. https://doi.org/10.1109/thfe2.1960.4503255

Lin, Patrick, and Fritz Allhoff. 2008. "Against Unrestricted Human Enhancement." *Journal of Evolution & Technology* 18 no.1: 35–41. https://doi.org/10.1007/s11569-008-0051-x

Lindstrom, Martin. 2005. *Brand Sense: Sensory Secrets Behind the Stuff We Buy*. London: Kogan Page.

Lipps, Theodor. 1979 [1903]. "Empathy, Inner Imitation and Sense-Feelings." In *A Modern Book of Esthetics*, edited by Melvin Rader, 374–382. New York: Holt, Rinehart and Winston.

Lister, Martin, Dovey, Jon, Giddings, Seth, Grant, Iain, and Kieran Kelly. 2003. *New Media: A Critical Introduction*. London: Routledge.

Liu, Yi-Ling. 2020. "The Future of the Classroom? China's Experience of AI in Education." *The AI Powered State: China's Approach to Public Sector Innovation*, 1–76.

Last modified April 22, 2022. https://media.nesta.org.uk/documents/Nesta_TheAIPo weredState_2020.pdf

Livingstone, Sonia, Mascheroni, Giovanna, and Elisabeth Staksrud. 2018. "European research on children's internet use: Assessing the past and anticipating the future." *New Media & Society* 20 no.3: 1103–1122. https://doi.org/10.1177/1461444816685930

Lomborg, Stine. 2022. "EVERYDAY AI AT WORK: Self-tracking and automated communication for smart work." In *Everyday Automation Experiencing and Anticipating Emerging Technologies*, edited by Sarah Pink, Martin Berg, Deborah Lupton, Minna Ruckenstein, 126–139. Abingdon: Routledge.

Luckhurst, Pheobe. 2020. "Priti Patel's Resting Smirk Face Makes Her Westminster's Perfect Pantomime Villain." *Vogue UK*. Last modified April 22, 2022. https://www. vogue.co.uk/news/article/priti-patel

Lupton, Deborah. 2018. "How do data come to matter? Living and becoming with personal data." *Big Data & Society* 5 no.2: https://doi.org/10.1177/2053951718786314

Man, Kingson, and Antonio Damasio. 2019. "Homeostasis and soft robotics in the design of feeling machines." *Nature Machine Intelligence* 1: 446–452. https://doi.org/ 10.1038/s42256-019-0103-7

Mann, Steve, Nolan, Jason, and Barry Wellman. 2003. "Sousveillance: inventing and using wearable computing devices for data collection in surveillance environments." *Surveillance & Society* 1 no.3: 331–355. https://doi.org/10.24908/ss.v1i3.3344

Maibom, Heidi Lene. 2007. "The Presence of Others." *Philosophical Studies: An International Journal for Philosophy in the Analytic Tradition* 123 no.2: 161–190.

Malgieri, Gianclaudio, and Marcello Ienca. 2021. "The EU regulates AI but forgets to protect our mind." *European Law Blog*. Last modified April 22, 2022. https://european lawblog.eu/2021/07/07/the-eu-regulates-ai-but-forgets-to-protect-our-mind/

Mantello, Peter, Ho, Manh-Tung, Nguyen, Minh-Hoang, and Quan-Hoang Vuong. 2021. "Bosses without a heart: Socio-demographic and cross-cultural determinants of attitude toward emotional ai in the workplace." *AI & Society*, 38: 1–23. https://doi. org/10.1007/s00146-021-01290-1

Maoz, Uri, Yaffe, Gideon, Koch, Christof, and Liad Mudrik. 2019. Neural precursors of decisions that matter-an ERP study of deliberate and arbitrary choice, *eLife*, 8. https:// doi.org/10.7554/elife.39787

Maoz, Uri, Ye, Shengxuan, Ross, Iain, Mamelak, Adam, and Christof Koch. 2012. "Predicting Action Content On-Line and in Real Time before Action Onset—an Intracranial Human Study." *Advances in Neural Information Processing Systems (NIPS)*. Last modified April 22, 2022. http://papers.nips.cc/paper/4513-predictingact ion-content-on-line-and-in-real-time-before-action-onset-an-intracranial-human-study.pdf

Marcus, Steven J. (ed.). 2002. *Neuroethics: Mapping the Field*. New York: Dana Press.

McCarthy, John, Minsky, Marvin, Rochester, Nathaniel, and Claude Shannon. 1955. "A Proposal for the Dartmouth Summer Research Project of Artificial Intelligence." Last modified April 22, 2022. http://jmc.stanford.edu/articles/dartmouth/dartmouth.pdf

McDuff, Daniel. .2014. "Crowdsourcing Affective Responses for Predicting Media Effectiveness." PhD Thesis. Last modified April 22, 2022. http://affect.media.mit.edu/ pdfs/14.McDuff-Thesis.pdf.

McDuff, Daniel, and Mary Czerwinski. 2018. "Designing emotionally sentient agents." *Communications of the ACM* 61 no.12: 74–83. http://dx.doi.org/10.1145/3186591

McDuff, Daniel, and Ashish Kapoor. 2019. "Visceral Machines: Risk-Aversion in Reinforcement Learning with Intrinsic Physiological Rewards." Last modified April 22, 2022. https://openreview.net/forum?id=SyNvti09KQ

McStay, Andrew. 2011. *The Mood of Information: A Critique of Behavioural Advertising.* New York: Continuum.

McStay, Andrew. 2013. *Creativity and Advertising: Affect, Events and Process.* London: Routledge.

McStay, Andrew. 2014. *Privacy and Philosophy: New Media and Affective Protocol.* New York: Peter Lang.

McStay, Andrew. 2016. "Empathic media and advertising: Industry, policy, legal and citizen perspectives (the case for intimacy)." *Big Data & Society* 3 no. 2. https://doi.org/10.1177/2053951716666868

McStay, Andrew. 2018. *Emotional AI: The Rise of Empathic Media.* London: Sage.

McStay, Andrew. 2019. "Emotional AI and EdTech: serving the public good." *Learning Media & Technology* 45 no. 3: 270–283. https://doi.org/10.1080/17439884.2020.1686016

McStay, . Andrew. 2021. "Emotional AI, Ethics, and Japanese Spice: Contributing Community, Wholeness, Sincerity, and Heart." *Philosophy & Technology* 34 no.4: 1781–1802. https://doi.org/10.1007/s13347-021-00487-y

McStay, Andrew. and Duncan Minty. 2019. *Emotional AI and Insurance: Online Targeting and Bias in Algorithmic Decision Making.* Last modified April 22, 2022 https://drive.google.com/file/d/15k_lmKB5BRGNZS1LqEJp2ZZV0zJ1wsQg/view,

McStay, Andrew. & Rosner, Gilad. 2021. "Emotional Artificial Intelligence in Children's Toys and Devices: Ethics, Governance and Practical Remedies." *Big Data & Society* 8 no.1. https://doi.org/10.1177/2053951721994877

McStay, Andrew. & Urquhart, Lachlan. 2019 "'This time with feeling?' Assessing EU data governance implications of out of home appraisal based Emotional AI." *First Monday* 24 no.10. https://doi.org/10.5210/fm.v24i10.9457

McStay, Andrew. & Urquhart, Lachlan. 2022. "In Cars (Are We Really Safest of All?): Interior Sensing and Emotional Opacity." *International Review of Law, Computers & Technology* 35 no.3: 1–24. https://doi.org/10.1080/13600869.2021.2009181

Mele, Alfred R. 2014a. *A Dialogue on Free Will and Science.* New York: Oxford University Press.

Mele, Alfred R. 2014b. "Free Will and Substance Dualism: The Real Scientific Threat to Free Will?" In *Moral Psychology, Volume 4*, edited by Walter Sinnott-Armstrong. 195–207. Cambridge, MA: MIT Press.

Mele, Vincenzo. 2015. "At the crossroad of Magic and Positivism'. Roots of an Evidential Paradigm through Benjamin and Adorno." *Journal of Classical Sociology* 15 no.2: 139–153. https://doi.org/10.1177/1468795X14567284

Mechant, Peter, De Wolf, Ralf, Van Compernolle, Mathias, Joris, Glen, Evens, Tom, and Lieven De Marez. 2021. "Saving the web by decentralizing data networks? A sociotechnical reflection on the promise of decentralization and personal data stores." *2021 14th CMI International Conference - Critical ICT Infrastructures and Platforms*, 1–6. https://doi.org/10.1109/cmi53512.2021.9663788

Mercedes-Benz. 2021. "How artificial intelligence is contributing to the car of the future." Last modified April 22, 2022. https://www.mercedes-benz.com.au/passengerc ars/experience/mercedes-me-magazine/innovation/articles/artificial-intelligence-car-of-future/stage.module.html

Michelfelder, Diane. 2015. "Postphenomenology with an Eye to the Future." In *Postphenomenological Investigations: Essays on Human-Technology Relations*, edited by Robert Rosenberger and Peter-Paul Verbeek, 237–246. London: Lexington Books.

Microsoft. 2021. "Person metrics." Last modified April 22, 2022. https://docs.microsoft.com/en-us/viva/insights/Use/Metric-definitions#person-metrics

Microsoft. 2022. "Face detection and attributes." Last modified April 22, 2022. https://docs.microsoft.com/en-gb/azure/cognitive-services/face/concepts/face-detection

Microsoft Education. 2022. "MR in your classroom." Last modified April 22, 2022. https://www.microsoft.com/en-us/education/mixed-reality

Mill, John S. 1962 [1859]. *Utilitarianism, On Liberty, Essay on Bentham*. London: Fontana Press.

Millett, David. 2001. "Hans Berger: From Psychic Energy to the EEG." *Perspectives in Biology and Medicine* 44 no.4: 522–542. https://doi.org/10.1353/pbm.2001.0070

Mirchandani, Kiran. 2003. "Challenging Racial Silences in Studies of Emotion Work: Contributions from Anti-Racist Feminist Theory." *Organization Studies* 24 no. 5: 721–742. https://doi.org/10.1177/0170840603024005003

Mirzoeff, Nicholas. 2020. "Artificial vision, white space and racial surveillance capitalism." *AI & Society* 36 no.4: 1295–1305. https://doi.org/10.1007/s00146-020-01095-8

Mitchell, Andrew J, and Peter Trawny. (2017) *Heidegger's Black Notebooks: Responses to Anti-Semitism*. Columbia University Press

Mittelstadt, Brent. 2017. "From Individual to Group Privacy in Big Data Analytics." *Philosophy & Technology* 30: 475–494. https://doi.org/10.1007/s13347-017-0253-7

Molenberghs, Pascal, and Winnifred R. Louis. 2018. "Insights from fMRI Studies Into Ingroup Bias." *Frontiers in Psychology* 9. https://doi.org/10.3389/fpsyg.2018.01868

Moore, Pheobe V. 2018. "Tracking Affective Labour for Agility in the Quantified Workplace." *Body & Society* 24 no.3: 39–67. https://doi.org/10.1177/1357034x18775203

Morgan, Kate. 2021. "Why in-person workers may be more likely to get promoted." BBC. Last modified April 22, 2022. https://www.bbc.com/worklife/article/20210305-why-in-person-workers-may-be-more-likely-to-get-promoted

Mulvin, Dylan. 2021. *Proxies: The Cultural Work of Standing In*. London: MIT.

Murali, Prasanth, Hernandez, Javier, McDuff, Daniel, Rowan, Kael, Suh, Jina, and Mary Czerwinski. 2021. "AffectiveSpotlight: Facilitating the Communication of Affective Responses from Audience Members during Online Presentations." In Proceedings of the 2021 CHI Conference on Human Factors in Computing Systems (CHI '21). *Association for Computing Machinery*. New York, NY, USA, Article 247, 1–13. https://doi.org/10.1145/3411764.3445235

Murata, Kiyoshi. 2019. "Japanese Traditional Vocational Ethics: Relevance and Meaning for the ICT- dependent Society." In *Tetsugaku Companion to Japanese Ethics and Technology*, edited by Thomas Taro Lennerfors & Kiyoshi Murata, 139–60. Berlin: Springer.

Muthukumar, Vidya, Pedapati, Tejaswini, Ratha, Nalini, Sattigeri, Prasanna, Wu, Chai-Wah, Kingsbury, Brian, Kumar, Abhishek, Thomas, Samuel, Mojsilovic, Aleksandra, and Kush R Varshney. 2018. "Understanding Unequal Gender Classification Accuracy from Face Images. *arXiv*, 1–11. https://arxiv.org/abs/1812.00099

Nagel, Thomas. 1986. *The View from Nowhere*. New York: OUP.

Narayanan, Arvind. 2021. "How to recognize AI snake oil." Last modified April 22, 2022. https://www.cs.princeton.edu/~arvindn/talks/MIT-STS-AI-snakeoil.pdf

NASUWT. 2021. "In the Classroom" Last modified April 22, 2022. https://www.nasuwt.org.uk/advice/in-the-classroom/class-sizes.html#

Needleman, Sarah E. (2020) "Judge Dismisses New Mexico Lawsuit Against Google Over Children's Data Privacy." *The Wall Street Journal*. Last modified April 22, 2022. https://www.wsj.com/articles/judge-dismisses-new-mexico-lawsuit-against-google-over-childrens-data-privacy-11601392392

Nelson, Ted. 1974 [2003]. "Computer Lib: You Can and Must Understand Computers Now." In *The New Media Reader*, edited by Nick Montfort and Noah Wardrip-Fruin, 308–316. Cambridge: MIT.

Neuralink. 2020. Neuralink Progress Update, Summer 2020, *YouTube*. Last modified April 22, 2022. https://www.youtube.com/watch?v=DVvmgjBL74w

Neuralink. 2022. "Interfacing with the Brain." Last modified April 22, 2022. https://neuralink.com/approach.

Nishida, Kitaro. 2003 [1958]. *Intelligibility and the Philosophy of Nothingness: Three Philosophical Essays*. Delhi: Facsimile.

Nishimoto, Shinji, Vu, An T, Naselaris, Thomas, Benjamini, Yuval, Yu, Bin, and Jack L. Gallant. 2011. "Reconstructing Visual Experiences from Brain Activity Evoked by Natural Movies." *Current Biology* 21 no. 19: 1641–1646. https://doi.org/10.1016/j.cub.2011.08.031

Nissenbaum, Helen. 2010. *Privacy in Context: Technology, Policy, and the Integrity of Social Life*. Stanford: Stanford University Press.

Noble, Safiya Umoja. 2018. *Algorithms of Oppression: How Search Engines Reinforce Racism*. New York: NYU Press.

Nozick, Robert. 2002 [1974]. *Anarchy, State, and Utopia*. Oxford: Blackwell.

Nussbaum, Martha. 2001. *Upheavals of Thought: The Intelligence of Emotions*. Cambridge University Press.

Obar, Jonathan A, and Anne Oeldorf-Hirsch. 2020. "The biggest lie on the Internet: ignoring the privacy policies and terms of service policies of social networking services." *Information, Communication & Society* 23 no.1=: 128–147. https://doi.org/10.1080/1369118x.2018.1486870

Oberman, Lindsay. and Ramachandran, Vilayanur S. 2009. "Reflections on the Mirror Neuron System: Their Evolutionary Functions Beyond Motor Representation." In *Mirror Neuron Systems: The Role of Mirroring Processes in Social Cognition*, edited by Jaime A. Pineda, 39–62. Totowa, NJ: Humana Press.

OECD. 2015. "Skills for Social Progress: The Power of Social and Emotional Skills." Last modified April 22, 2022. https://doi.org/10.1787/9789264226159-en.

OECD. 2018. "World Class How to Build a 21st-Century School System." Last modified April 22, 2022. http://www.oecd.org/education/world-class-9789264300002-en.htm.

OECD. 2019. "Recommendation of the Council on Responsible Innovation in Neurotechnology." Last modified April 22, 2022. https://legalinstruments.oecd.org/api/print?ids=658&Lang=en

OHCHR. 2021. "OHCHR and privacy in the digital age." Last modified April 22, 2022. https://www.ohchr.org/en/issues/digitalage/pages/digitalageindex.aspx

O'Malley, Pat. 1996. Risk and responsibility. In *Foucault and Political Reason*, edited by Andrew Barry, Thomas Osborne, and Nikolas Rose, 189–207. Chicago, IL: University of Chicago Press.

Osher, David, Kidron, Yael, Brackett, Marc, Dymnicki, Allison, Jones, Stephanie, and Roger P. Weissberg. 2016. "Advancing the Science and Practice of Social and Emotional Learning: Looking Back and Moving Forward." *Review of Research in Education* 40: 644–681. https://doi.org/10.3102/0091732x16673595

Ott, M. 2020. "Affection and Dividuation." In *Affective Transformations: Poliitcs, Algorithms, Media*, edited by Bernd Bösel and Serjoscha Wiemer, 187–199. Meson Press.

Pangrazio, Luci and Selwyn, Neil. 2019. "'Personal data literacies': A critical literacies approach to enhancing understandings of personal digital data." *New Media & Society* 21 no.2: 419–437. https://doi.org/10.1177/1461444818799523

Parcollet, Titouan, and Ravanelli, Mirco. 2021. The Energy and Carbon Footprint of Training End-to-End Speech Recognizers. *INTERSPEECH*, 4583–4587. https://doi.org/10.21437/interspeech.2021-456.

Parker, K. 2021. "The undeniable benefits of video interviewing - critics of VI are selling you snake oil." Last modified April 22, 2022. https://www.hirevue.com/blog/hiring/the-undeniable-benefits-of-video-interviewing-critics-of-vi-are-selling-you-snake-oil

Parisi, Domenico. 2004. "Internal robotics." *Connection science* 16 no.4: 325–338. https://doi.org/10.1080/09540090412331314768

Pentland, Alex, and Thomas Hardjono. 2020. "Data Cooperatives." Last modified April 22, 2022. https://wip.mitpress.mit.edu/pub/pnxgvubq/release/2

Peters, Michael A. 2016. "Inside the Global Teaching Machine: Moocs, Academic Labour and the Future of The University." *Learning and Teaching* 9 no.2: 66–88. https://doi.org/10.3167/latiss.2016.090204

Pfeifer Rolf, and Fumiya Iida. 2004. "Embodied Artificial Intelligence: Trends and Challenges." In: *Embodied Artificial Intelligence. Lecture Notes in Computer Science, vol 3139*. Fumiya Iida, Rolf Pfeifer, Luc Steels, and Yasuo Kuniyoshi, 1–26. Berlin: Springer.

Philippopoulos-Mihalopoulos, Andreas. 2015. *Spatial Justice: Body, Lawscape, Atmosphere*. Abingdon: Routledge.

Picard, Rosalind W. 1997. *Affective Computing*. Cambridge, MA: MIT.

Picard, Rosalind W, Cassell, Justine, Kort, Barry, Reilly, Rob, Bickmore, Timothy, Kapoor, Ashish, Mota, Selene, and Catherine Vaucelle. 2001. "Affective Learning Companion." Last modified April 22, 2022. https://affect.media.mit.edu/projectpages/lc/.

Picard, Rosalind W, Kort, Barry, and Rob Reilly. 2001. "Exploring the Role of Emotion in Propelling the SMET Learning Process." Last modified April 22, 2022. https://affect.media.mit.edu/projectpages/lc/nsf1.PDF.

Pinghui, Zhuang. 2017. "Chinese Women Offered Refund After Facial Recognition Allows Colleague to Unlock iPhone X." *South China Morning Post*. Last modified April 22, 2022. https://www.scmp.com/news/china/society/article/2124313/chinese-woman-offered-refund-after-facial-recognition-allows

Pink, Sarah, Berg, Martin,. Lupton, Deborah, and Minna Ruckenstein. 2022. "Introduction: Everyday Automation: setting a research agenda." In *Everyday Automation Experiencing and Anticipating Emerging Technologies*, edited by Sarah Pink, Martin Berg, Deborah Lupton, and Minna Ruckenstein, 1–19. Abingdon: Routledge.

Plato (1987 [375 BC]) *The Republic*. London: Penguin.

Potamias, Rolandos Alexandros, Siolas, Siolas, and Andreas-Georgios Stafylopatis. 2020. "A transformer-based approach to irony and sarcasm detection." *Neural Computing and Applications*, 32: 17309–17320. https://doi.org/10.1007/s00521-020-05102-3

Proctortrack. 2022. "How it works." Last modified April 22, 2022. https://www.proctortrack.com/how-it-works/

Pugliese, Joseph. 2007. "Biometrics, infrastructural whiteness, and the racialized degree zero of nonrepresentation." *Boundary 2* 34 no. 2: 105–133. https://doi.org/10.1215/01903659-2007-005

Quetelet, Adolphe. 1842. *A Treatise on man and the Development of his Faculties*. Edinburgh: William and Robert Chambers.

Rahm, Lina, and Anne Kaun. 2022. "Imagining mundane automation: Historical trajectories of meaning-making around technological change." In *Everyday Automation Experiencing and Anticipating Emerging Technologies*, edited by Sarah Pink, Martin Berg, Deborah Lupton, and Minna Ruckenstein, 23–43. Abingdon: Routledge.

Rawls, John. 1971. *A Theory of Justice*. Cambridge: Harvard University Press.

Rawls, John. 1993. *Political Liberalism*, New York: Columbia University Press.

Rhue, Lauren. 2018. "Racial Influence on Automated Perceptions of Emotions." *SSRN*, 1–11. https://papers.ssrn.com/sol3/papers.cfm?abstract_id=3281765.

Rhue, Lauren. 2019. "Emotion-reading tech fails the racial bias test." *The Conversation*. Last modified April 22, 2022. https://theconversation.com/emotion-reading-tech-fails-the-racial-bias-test-108404

Roach, John. 2021. "Mesh for Microsoft Teams aims to make collaboration in the 'metaverse' personal and fun." Last modified April 22, 2022. https://news.microsoft.com/innovation-stories/mesh-for-microsoft-teams/

Romm, Tony. 2020. "Tech Giants Led by Amazon, Facebook and Google Spent Nearly Half a Billion on Lobbying over the Past Decade, New Data Shows." *The Washington Post*. Last modified June 30, 2023. https://www.washingtonpost.com/technology/2020/01/22/amazon-facebook-google-lobbying-2019/

Rorty, Richard. 2007. *Philosophy as Cultural Politics: Philosophical Papers 4*. Cambridge: Cambridge University Press.

Rucker, Rudy, Sirius, R.U, and Queen Mu (eds). 1993. *Mondo 2000: A User's Guide to the New Edge*. London: Thames and Hudson.

Ruhaak, Anouk. 2019. "Data Trusts: Why, What and How." Last modified April 22, 2022. https://medium.com/@anoukruhaak/data-trusts-why-what-and-how-a8b53b53d34

Russell, James A. 1994. "Is there universal recognition of emotion from facial expression? A review of the cross-cultural studies." *Psychological Bulletin*, 115 no.1: 102–41. https://doi.org/10.1037/0033-2909.115.1.102

Russell, James A. 2003. "Core affect and the psychological construction of emotion." *Psychological Review* 110: 145–172. https://doi.org/10.1037/0033-295x.110.1.145

Ryle, Gilbert. 2000 [1949]. *The Concept of Mind*. Penguin: London.

SAE International. 2018. "SAE International Releases Updated Visual Chart for Its "Levels of Driving Automation" Standard for Self-Driving Vehicles." Last modified April 22, 2022. https://www.sae.org/news/press-room/2018/12/sae-internatio nal-releases-updated-visual-chart-for-its-%E2%80%9Clevels-of-driving-automat ion%E2%80%9D-standard-for-self-driving-vehicles

Salvaris, Mathew, and Patrick Haggard. 2014. "Decoding intention at sensorimotor timescales." *PLoS ONE* 9 no.2: 1–11. http://doi.org/10.1371/journal.pone.0085100

Sadowski, Jathan. 2019. "When data is capital: Datafication, accumulation, and extraction." *Big Data & Society* 6 no.1: 1–12. https://doi.org/10.1177/2053951718820549

Javier Sánchez-Monedero, Lina Dencik, and Lilian Edwards. 2020. What does it mean to 'solve' the problem of discrimination in hiring? Social, technical and legal perspectives from the UK on automated hiring systems. In Conference on Fairness, Accountability, and Transparency (FAT* '20), January 27–30, 2020, Barcelona, Spain. ACM, New York, NY, USA, 11 pages. https://doi.org/10.1145/3351095.3372849

Sandel, Michael. 2009. *Justice: What's The Right Thing To Do?*. London: Penguin.

Sandel, Michael. 2012. *What Money Can't Buy: The Moral Limits of Markets*. London: Penguin.

Satel, Sally, and Scott O. Lilienfeld. 2013. *Brainwashed: The Seductive Appeal of Mindless Neuroscience*. New York: Basic Books.

Scarantino, Andrea. 2012. "How to Define Emotions Scientifically." *Emotion Review* 4 no.4: 358–368. https://doi.org/10.1177/1754073912445810

Scarantino, Andrea. 2018. "The Philosophy of Emotions and its Impact on Affective Science." In *Handbook of Emotions* (4th Edition), edited by Lisa Feldman Barrett, Michael Lewis, Jeanette Haviland-Jones, 3–48. New York: Guildford.

Scarff, Robbie. 2021. "Emotional Artificial Intelligence, Emotional Surveillance, and the Right to Freedom of Thought." Last modified April 22, 2022. https://easychair.org/publications/preprint_open/qJfZ

Schiller, Devon. 2021. "The face and the faceness: Iconicity in the early faciasemiotics of Paul Ekman, 1957–1978." *Sign Systems Studies* 49 no.3/4: 361–382. https://doi.org/10.12697/sss.2021.49.3-4.06

Schuller, Bjorn W. 2017. "Editorial: IEEE Transactions on Affective Computing—Challenges and Chances." IEEE Transactions on Affective Computing 8 no.1: 1–2. https://doi.org/10.1109/ TAFFC.2017.2662858.

Schopenhauer, Arthur. 1960 [1841]. *Essay on the Freedom of the Will*. Indianapolis: The Liberal Arts Press/Bobbs Merrill.

Schwartz, Roy, Dodge, Jesse, Smith, Noah A, and Oren Etzioni. 2020. "Green AI." *Communications of the ACM* 63 no.12: 54–63. https://doi.org/10.1145/3381831

Sebeok, Thomas A. 1975. The semiotic web: A chronicle of prejudices. Bulletin of Literary Semiotics 2: 1–63. https://doi.org/10.5840/bls197521

Selwyn, Neil. 2017. *Education and Technology: Key Issues and Debates.* London: Bloomsbury.

Selwyn, Neil. 2021. "Ed-Tech Within Limits: Anticipating educational technology in times of environmental crisis." *E-Learning and Digital Media* 18 no.5: 496–510. https://doi.org/10.1177/20427530211022951

Sententia, Wrye. 2004. "Neuroethical Considerations: Cognitive Liberty and Converging Technologies for Improving Human Cognition." *Annals of the New York Academy of Sciences* 1013 no.1: 221–8. https://doi.org/10.1196/annals.1305.014

Shannon, Claude. E. 1976 [1938]. "A Symbolic Analysis of Relay and Switching Circuits." In *Claude Elwood Shannon: Collected Papers*, edited by N.J.A. Sloane and Aaron D. Wyner, 471–495. New York: IEEE Press.

Shannon, Claude E, and Warren Weaver. 1949 *The Mathematical Theory of Communication.* Illinois: University of Illinois Press.

Sharp Eleni. 2021. "Personal data stores: building and trialling trusted data services." *BBC Research & Development.* Last modified April 22, 2022. https://www.bbc.co.uk/rd/blog/2021-09-personal-data-store-research.

Shaver, Phillip, Schwartz, Judith, Kirson, Donald, and Cary O'Connor. 1987. "Emotion knowledge: further exploration of a prototype approach." *Journal of Personality and Social Psychology* 52 no.6: 1061–1086. https://doi.org/10.1037/0022-3514.52.6.1061

Sheller, Mimi. 2004. "Automotive Emotions: Feeling the Car." *Theory, Culture & Society* 21 no.4–5: 221–242. https://doi.org/10.1177/0263276404046068

Shu, Catherine. 2020. "Chinese online education app Zuoyebang raises $1.6 billion from investors including Alibaba." Last modified April 22, 2022. https://techcrunch.com/2020/12/28/chinese-online-education-app-zuoyebang-raises-1-6-billion-from-investors-including-alibaba/

Shuster, Anastasia, Inzelberg, Lilah, Ossmy, Ori, Izakson, Liz, Hanein, Yael, and Levy Dino. 2021. "Lie to my face: An electromyography approach to the study of deceptive behaviour." *Brain and Behavior* 11 no.12: 1–12. https://doi.org/10.1002/brb3.2386

Silverstone, Roger. 2002. "Complicity and Collusion in the Mediation of Everyday Life." *New Literary History* 33 no.4: 745–764. https://doi.org/10.1353/nlh.2002.0045

Singer, Natasha. 2021. "Learning Apps Have Boomed in the Pandemic. Now Comes the Real Test." *The New York Times.* Last modified April 22, 2022. https://www.nytimes.com/2021/03/17/technology/learning-apps-students.html

Smith, Adam. 2011 [1759]. *The Theory of Moral Sentiments.* Kapaau: Gutenberg.

Smith, Tiffany W. 2015. *The Book of Human Emotions.* London: Profile.

Smythe, Dallas W. (1981) *Dependency Road: Communications, Capitalism, Consciousness, and Canada.* New Jersey: ABLEX.

Soon, Chun Siong, Brass, Marcel, Heinze, Hans-Jochen, and John-Dylan Haynes. 2008. "Unconscious determinants of free decisions in the human brain." *Nature Neuroscience* 11 no.5: 543–545. https://doi.org/10.1038/nn.211

Spengler, Oswald. 1926. *The Decline of the West: Form and Actuality.* New York: Alfred A. Knopf.

Spiekermann, Sarah, Acquisti, Alessandro, Böhme, Rainer, and Kai-Lung Hui. 2015. "The Challenges of Personal Data Markets and Privacy." *Electronic Markets* 25 no.2: 161–167. https://doi.org/10.1007/s12525-015-0191-0

Spinoza, Baruch. 1996 [1677]. *Ethics*. London: Penguin.

Stark, Luke. 2020. "The emotive politics of digital mood tracking." *New Media & Society* 22 no.11: 2039–2057. https://doi.org/10.1177/1461444820924624

Stark, Luke, and Jesse Hoey. 2020. *The Ethics of Emotion in Artificial Intelligence Systems*, OSFPreprint, 1–12. https://doi.org/10.31219/osf.io/9ad4u

Stark, Luke, and Hutson, Jevan. (2021) Physiognomic Artificial Intelligence, SSRN, 1–39. https://doi.org/10.2139/ssrn.3927300

Stearns, Peter N, and Carol Z. Stearns. 1985. "Emotionology: Clarifying the History of Emotions and Emotional Standards." *The American Historical Review* 90 no. 4: 813–836. https://doi.org/10.2307/1858841

Straiton, Jenny, and Francesca Lake. 2021. "Inside the brain of a killer: the ethics of neuroimaging in a criminal conviction" *Biotechniques*, 70 no.2: 69–71. https://doi.org/10.2144/btn-2020-0171

Steedman, Robin, Kennedy, Helen, and Rhianne Jones. 2020. "Complex ecologies of trust in data practices and data-driven systems." *Information, Communication and Society* 23 no.6: 817–832. https://doi.org/10.1080/1369118x.2020.1748090

Stenner, Paul. 2020. "Affect: On the Turn." In *Affective Transformations: Politics, Algorithms, Media*, edited by Bernd Bösel and Serjoscha Wiemer, 19–39. Meson Press.

Suchman, Lucy. 2007. *Human-Machine Reconfigurations: Plans and Situated Actions*. Cambridge: Cambridge University Press.

Taylor, Linnet. 2017. "What is data justice? The case for connecting digital rights and freedoms globally." *Big Data & Society* 4 no.2: 1–14. https://doi.org/10.1177/2053951717736335

Tech at Meta. 2021a. "Reality Labs." Last modified April 22, 2022. https://tech.fb.com/ar-vr/

Tech at Meta. 2021b. "Inside Facebook Reality Labs: Wrist-based interaction for the next computing platform." Last modified April 22, 2022. https://tech.fb.com/ar-vr/2021/03/inside-facebook-reality-labs-wrist-based-interaction-for-the-next-computing-platform/

The AI Index. 2021. "The AI Index 2021 Annual Report by Stanford University." Last modified April 22, 2022 https://aiindex.stanford.edu/wp-content/uploads/2021/03/2021-AI-Index-Report_Master.pdf

Thompson, Courtney E. 2021. *An Organ of Murder Crime, Violence, and Phrenology in Nineteenth-Century America*. NJ: Rutgers University Press

Thompson, Neil C, Greenwald, Kristjan, Lee, Keeheon, and Gabriel F. Manso, 2021. DEEP LEARNING'S DIMINISHING RETURNS, *IEEE Spectrum* 58 no.10: 50–55. https://doi.org/10.1109/mspec.2021.9563954

Thomson, Judith Jarvis. 1985. "The Trolley Problem." *The Yale Law Journal* 94 no.6: 1395–1415. https://doi.org/10.2307/796133

Tomkins, Silvan S. 1962. *Affect, imagery, consciousness: Vol. 1. The positive affects*. New York: Springer.

Torres, Émile P. 2022. "Understanding "longtermism": Why this suddenly influential philosophy is so toxic." *Salon*. Last modified Feb 03, 2023. https://www.salon.com/2022/08/20/understanding-longtermism-why-this-suddenly-influential-philosophy-is-so/

TRIMIS. 2021. "Vision Zero Initiative." Last modified April 22, 2022 https://trimis. ec.europa.eu/?q=project/vision-zero-initiative#tab-outline

TUC. 2020. "Technology managing people: The worker experience." Last modified April 22, 2022. https://www.tuc.org.uk/sites/default/files/2020-11/Technology_Managing_ People_Report_2020_AW_Optimised.pdf

TUC. 2021. "WHEN AI IS THE BOSS AN INTRODUCTION FOR UNION REPS." Last modified April 22, 2022. https://www.tuc.org.uk/sites/default/files/When_AI_ Is_The_Boss_2021_Reps_Guide_AW_Accessible.pdf

Tucker, Mary Evelyn. 1998. "Religious dimensions of Confucianism: Cosmology and cultivation." *Philosophy East and West,* 48 no.1: 5–45. https://doi.org/10.1007/978-3-319-59027-1_5

Turing, Alan M. (1950) Computing Machinery and Intelligence, *Mind*, 49: 433–460.

Turkle, Sherry. 1984 [2005]. *The Second Self: Computers and the Human Spirit.* Cambridge, MA: MIT.

Turkle, Sherry. 2019. Faking Intimacy, *New Philosopher*, 22: 21–22.

UNCRC. 2013. "General comment No. 14 (2013) on the right of the child to have his or her best interests taken as a primary consideration (art. 3, para. 1)." Last modified April 22, 2022. https://www2.ohchr.org/english/bodies/crc/docs/gc/crc_c_gc_ 14_eng.pdf

UNCRC. 2021. "General comment No. 25 (2021) on children's rights in relation to the digital environment." Last modified April 22, 2022. https://docstore.ohchr.org/SelfS ervices/FilesHandler.ashx?enc=6QkGld%2FPPRiCAqhKb7yhsqIkirKQZLK2M5 8RF%2F5F0vEG%2BcAAx34gC78FwvnmZXGFUl9nJBDpKRldfKekJxW2w9nNry RsgArkTJgKelqeZwK9WXzMkZRZd37nLNlbFc2t

UNESCO. 2021a. "Reimagining our futures together: a new social contract for education." Last modified April 22, 2022. https://reliefweb.int/report/world/reimagining-our-futures-together-new-social-contract-education

UNESCO. 2021b. "DRAFT RECOMMENDATION ON THE ETHICS OF ARTIFICIAL INTELLIGENCE." Last modified April 22, 2022. https://unesdoc.unesco.org/ark:/ 48223/pf0000378931/PDF/378931eng.pdf.multi

UN General Assembly. 2021. "Resolution adopted by the Human Rights Council on 7 October 2021." Last modified April 22, 2022. https://undocs.org/A/HRC/RES/48/4

United States Patent and Trademark Office. 2020. "Meeting Insight Computing System." Last modified April 22, 2022. *US020200358627A120201112*, https://pdfaiw.uspto. gov/.aiw?PageNum=0&docid=20200358627&IDKey=19C7DC5929AB&Home Url=http%3A%2F%2Fappft.uspto.gov%2Fnetacgi%2Fnph-Parser%3FSect1%3DP TO1%2526Sect2%3DHITOFF%2526d%3DPG01%2526p%3D1%2526u%3D%252 52Fnetahtml%25252FPTO%25252Fsrchnum.html%2526r%3D1%2526f%3D G%25261%3D50%2526s1%3D%2525222020 0358627%252522.PGNR.%252 6OS%3DDN%2F20200358627%2526RS%3DDN%2F20200358627

Urquhart, Lachlan, Sailaja, Neelima, and Derek McAuley. 2018. Realising the right to data portability for the domestic Internet of things. Personal and Ubiquitous Computing, 22: 317–332. https://doi.org/10.1007/s00779-017-1069-2

van Wynsberghe, Aimee. (2021). Sustainable AI: AI *for* sustainability and the sustainability *of* AI. *AI and Ethics*, 1 no.3: 213–218. https://doi.org/10.1007/s43681-021-00043-6

Varela, Francisco J. (1996) NEUROPHENOMENOLOGY A Methodological Remedy for the Hard Problem, *Journal of Consciousness Studies*, 3 no.4: 330–349. https://unsta ble.nl/andreas/ai/langcog/part3/varela_npmrhp.pdf

Varshizky, Amit. 2021. "Non-Mechanistic Explanatory Styles in the German Racial Science: A Comparison of Hans F. K. Günther and Ludwig Ferdinand Clauß." In *Recognizing the Past in the Present*, edited by Sabine Hildebrandt, Miriam Offer, and Michael A. Grodin, 21–43. New York: Berghahn.

Vattimo, Gianni. 1988. *The End of Modernity*. Cambridge: Polity Press.

Velasco, Emily. 2021. "Recording Brain Activity with Laser Light." Last modified April 22, 2022. https://www.caltech.edu/about/news/recording-brain-activity-with-laser-light

Verbeek, Peter-Paul. 2011a. *Moralizing Technology: Understanding and Designing the Morality of Things*. Chicago: University of Chicago Press.

Verbeek, Peter-Paul. 2011b. "Subject to technology." In *Law, Human Agency, and Autonomic Computing: The Philosophy of Law Meets the Philosophy of Technology*, edited by Mireille Hildebrandt and Antoinette Rouvroy, 27–45. Abingdon: Routledge.

Vidal, Jacques J. 1973. "Toward Direct Brain-Computer Communication." Last modified April 22, 2022. https://www.caltech.edu/about/news/recording-brain-activity-with-laser-light   https://www.annualreviews.org/doi/pdf/10.1146/annurev.bb.02.060 173.001105

Viljoen, Salomé. 2020. "Data as Property?" *Phenomenal World*. Last modified April 22, 2022. https://phenomenalworld.org/analysis/data-as-property

Vincent, James. 2022. "Google is using a new way to measure skin tones to make search results more inclusive." *The Verge*. Last modified May 12, 2022. https://www.theverge.com/2022/5/11/23064883/google-ai-skin-tone-measure-monk-scale-inclusive-sea rch-results

Wallach, Wendell, and Colin Allen. 2008. *Moral machines: Teaching robots right from wrong*. Oxford: Oxford University Press.

Waller, Robyn Repko. 2019. Recent Work on Agency, Freedom, and Responsibility: A Review. Last modified April 22, 2022. https://www.templeton.org/wp-content/uplo ads/2019/10/Free-Will-White-Paper.pdf

Watsuji, Tetsurō [1937] 1996. *Watsuji Tetsurō's Rinrigaku: Ethics in Japan*. New York: SUNY Press.

Watsuji, Tetsurō. 1961. *A Climate: A Philosophical Study*. Printing Bureau: Japanese Government.

Watters, Audrey. 2014. "Teaching Machines: A Brief History of "Teaching at Scale." Last modified April 22, 2022. http://hackeducation.com/2014/09/10/teaching-machines-teaching-at-scale.

WEF. 2016. "New Vision for Education: Fostering Social and Emotional Learning through Technology." Last modified April 22, 2022. http://www3.weforum.org/docs/WEF_New_Vision_for_Education.pdf.

Whitehead, Mark, Jones, Rhys, Lilley, Rachel, Pykett, Jessica, and Rachel Howell. 2018. *Neuroliberalism: behavioural government in the twenty-first century*. London: Routledge.

Whittaker, Meredith. 2021 "The steep cost of capture." *Interactions* 28 no.6: 50–55. https://doi.org/10.1145/3488666

WHO. 2016. "Road safety." Last modified April 22, 2022. https://www.who.int/data/gho/data/themes/road-safety

WHO. 2022. "Decade of Action for Road Safety (2021–2030)." Last modified April 22, 2022. https://www.who.int/teams/social-determinants-of-health/safety-and-mobility/decade-of-action-for-road-safety-2021-2030

Wiemer, Serjoscha. (2020) Happy, Happy, Sad, Sad: Do You Feel Me? Constellations of Desires in Affective Technologies. In *Affective Transformations: Politics, Algorithms, Media*, edited by Bernd Bösel and Serjoscha Wiemer, 153–168. Meson Press.

Wiener, Norbert. 1985 [1948]. *Cybernetics or control and communication in the animal and the machine.* Cambridge, MA: MIT Press.

Wiener, Norbert. 1960. "Some Moral and Technical Consequences of Automation." *Science* 131 no.3410: 1355–1358. https://doi.org/10.1126/science.131.3410.1355

Williamson, Ben. 2018. Learning from psychographic personality profiling, Code Acts in Education. https://codeactsineducation.wordpress.com/2018/03/23/learning-from-psychographics

Williamson, Ben. 2021a. "Psychodata: disassembling the psychological, economic, and statistical infrastructure of 'social-emotional learning." *Journal of Education Policy* 36 no.1: 129–154. https://doi.org/10.1080/02680939.2019.1672895

Williamson, Ben. 2021b. Google's plans to bring AI to education make its dominance in classrooms more alarming, *Fast Company*, https://www.fastcompany.com/90641049/google-education-classroom-ai

Wilson, Benjamin, Hoffman, Judy, and Jamie Morgenstern. 2019. "Predictive Inequity in Object Detection." *arXiv*, 1–11. https://arxiv.org/abs/1902.11097

Winner, Langdon. 1986. *The Whale and the Reactor: A Search for Limits in an Age of High Technology.* Chicago: University of Chicago Press.

Wong, Pak-Hang. 2012. "Dao, harmony and personhood: Towards a Confucian ethics of technology." *Philosophy and Technology* 25 no.1: 67–86. https://doi.org/10.1007/s13347-011-0021-z

Wu, Xiaolin, and Xi Zhang. 2016. "Automated Inference on Criminality using Face Images." *arXiv*, 1–9. https://arxiv.org/pdf/1611.04135v1.pdf.

Yourdon, Edward. 1972. *Design of On-Line Computer Systems.* Englewood Cliffs, N.J.: Prentice-Hall.

Yū, Inutsuka. (2019) "A Moral Ground for Technology: Heidegger, Postphenomenology, and Watsuji." In *Tetsugaku Companion to Japanese Ethics and* Technology, edited by Thomas Taro Lennerfors & Kiyoshi Murata, 41–56. Berlin: Springer.

Yuste, Rafael, Goering, Sara, Agüera y Arcas, Blaise, Bi, Guoqiang, Carmena, Jose M, Carter, Adrian, and Joseph J. Fins, et al. 2017. "Four Ethical Priorities for Neurotechnologies and AI." *Nature* 551 no.7679: 159–163. https://doi.org/10.1038/551159a.

Zepf, Sebastian, Dittrich, Monique, Hernandez, Javier, and Alexander Schmitt. 2019. "Towards Empathetic Car Interfaces: Emotional Triggers while Driving." In *Extended Abstracts of the 2019 CHI Conference on Human Factors in Computing Systems (CHI EA '19).* Association for Computing Machinery, 1–6. NY, Paper LBW0129. https://doi.org/10.1145/3290607.3312883.

Zepf, Sebastian, Hernandez, Javier, Schmitt, Alexander, Minker, Wolfgang, and Rosalind W. Picard. 2020. "Driver Emotion Recognition for Intelligent Vehicles: A Survey." *ACM Computing Surveys* 53 no.3: 1–30. https://doi.org/10.1145/3388790.

Zelevansky, Nora, 2019. "The Big Business of Unconscious Bias." *The New York Times*. Last modified April 22, 2022. https://www.nytimes.com/2019/11/20/style/diversity-consultants.html.

Zimmerman, Michael E. 1990. *Heidegger's Confrontation with Modernity: Technology, Politics, Art*. Bloomington, Indianapolis: Indiana University Press.

Zuboff, Shoshana. 2019. *The Age of Surveillance Capitalism: The Fight for a Human Future at the New Frontier of Power*. London: Profile.

For the benefit of digital users, indexed terms that span two pages (e.g., 52–53) may, on occasion, appear on only one of those pages.
Tables are indicated by *t* following the page number